This valuable volume exposes the failure of current institutional arrangements, in the form of cultural outlooks, governance arrangements, democratic consensus, purposeful regulation, and even innovation in technology and human behaviour, to come to terms with the overwhelming barriers to achieving the ambition of the Paris Agreement to remove the planet of dangerous climate change within three preciously short generations. It should be read with care and attention as it sets the marker for the scale of institutional reform that surely will be required in our lifetimes.

Tim O'Riordan, *Emeritus Professor of Environmental Sciences at the University of East Anglia*

The Paris Climate Agreement signals a fundamental change in policy architecture. This book offers much needed ammunition against too easy attribution of catastrophic events to global warming – overlooking the importance of institutional and political contexts – and too easy confinement of 'solutions' to individual actions or markets – overlooking the role of multi-level, polycentric responses.

Arthur C. Petersen, *Professor of Science, Technology and Public Policy, University College London*

Anthropogenic climate change is a global challenge that also needs to be tackled through decisions made at the local/national level. To date, little knowledge has accumulated on how and when different local/national institutional frameworks are mobilized to address global challenges. *Institutional Capacity for Climate Change Response* thought-provokingly addresses this knowledge gap.

Katherine Richardson, *Professor and Leader of the Sustainability Science Centre, University of Copenhagen*

For decades, Steve Rayner was part of a band of academics and practitioners who – heretically, at the time – argued that the Kyoto Protocol would not be fully implemented or renewed, as attempting to do so would far exceed the capacities of local, national and international institutions. Now that this one-time heresy has quietly become dogma in recent years, we should pay close attention to what he and his colleagues have to say on how to strengthen the institutions responsible for climate change adaptation and mitigation. That is the theme of this excellent volume. Importantly, the book punctures the myth that authoritarian regimes are necessarily 'better' at combating climate change, while it also provides valuable insights into how democratic institutions can be strengthened and utilised better. I strongly recommend this book to all those interested in effectively addressing climate change now that the Kyoto Protocol has collapsed.

Marco Verweij, *Professor of Political Science, Jacobs University Bremen*

T0179049

Institutional Capacity for Climate Change Response

In a period of rapid climate change and climate governance failures, it is crucial to understand and address how effectively different political institutions can and should react to climate change.

The term 'institutional response capacity' can be defined as a measurement for how effective political institutions may respond to threats and challenges such as climate change. This book sets out to provide a venue for the discussion of how to conduct climate politics by offering new perspectives on how social and political institutions are capable of responding to climate change. In doing so, the book explores how democracy, institutional design and polycentric governance influence social and political entities' capacity to mitigate, adapt, address and transform climate change. The book offers building blocks for a new agenda of climate studies by focusing on institutional response capacity and by offering a new approach to climate governance at a time when many political initiatives have failed.

This interdisciplinary volume is a valuable resource for academics, researchers and policy-makers in the areas of anthropology, political science, geography and environmental studies.

Theresa Scavenius is an associate professor at the Department of Planning, University of Aalborg, Copenhagen, Denmark.

Steve Rayner is James Martin Professor of Science and Civilisation and Director of the Institute for Science, Innovation, and Society, University of Oxford, UK.

Science in Society Series
Series Editor: Steve Rayner
Institute for Science, Innovation and Society, University of Oxford

Editorial Board: Jason Blackstock, Bjorn Ola Linner, Susan Owens, Timothy O'Riordan, Arthur Petersen, Nick Pidgeon, Dan Sarewitz, Andy Stirling, Chris Tyler, Andrew Webster, Steve Yearley

The Earthscan Science in Society Series aims to publish new high-quality research, teaching, practical and policy-related books on topics that address the complex and vitally important interface between science and society.

Experiment Earth
Responsible innovation in geoengineering
Jack Stilgoe

Rethinking Greenland and the Arctic in the Era of Climate Change
New Northern Horizons
Frank Sejersen

International Science and Technology Education
Exploring Culture, Economy and Social Perceptions
Edited by Ortwin Renn, Nicole C. Karafyllis, Andreas Hohlt and Dorothea Taube

Policy Legitimacy, Science and Political Authority
Knowledge and action in liberal democracies
Edited by Michael Heazle and John Kane

Systems Thinking for Geoengineering Policy
How to reduce the threat of dangerous climate change by embracing uncertainty and failure
Robert Chris

Assessing the Societal Implications of Emerging Technologies
Anticipatory governance in practice
Evan Michelson

Aid, Technology and Development
The lessons from Nepal
Edited by Dipak Gyawali, Michael Thompson and Marco Verweij

Climate Adaptation Policy and Evidence
Understanding the Tensions between Politics and Expertise in Public Policy
Peter Tangney

Cities and the Knowledge Economy
Promise, Politics and Possibilities
Tim May and Beth Perry

Institutional Capacity for Climate Change Response
A New Approach to Climate Politics
Theresa Scavenius and Steve Rayner

Institutional Capacity for Climate Change Response

A New Approach to Climate Politics

Edited by Theresa Scavenius and Steve Rayner

LONDON AND NEW YORK

First published 2018 by Routledge

2 Park Square, Milton Park, Abingdon, Oxfordshire OX14 4RN

52 Vanderbilt Avenue, New York, NY 10017

First issued in paperback 2019

Routledge is an imprint of the Taylor & Francis Group, an informa business

© 2018 selection and editorial matter, Theresa Scavenius and Steve Rayner; individual chapters, the contributors

The right of Theresa Scavenius and Steve Rayner to be identified as the authors of the editorial matter, and of the contributors for their individual chapters, has been asserted in accordance with sections 77 and 78 of the Copyright, Designs and Patents Act 1988.

All rights reserved. No part of this book may be reprinted or reproduced or utilised in any form or by any electronic, mechanical, or other means, now known or hereafter invented, including photocopying and recording, or in any information storage or retrieval system, without permission in writing from the publishers.

Trademark notice: Product or corporate names may be trademarks or registered trademarks, and are used only for identification and explanation without intent to infringe.

British Library Cataloguing in Publication Data
A catalogue record for this book is available from the British Library

Library of Congress Cataloging in Publication Data
Names: Scavenius, Theresa, editor. | Rayner, Stephen, editor.
Title: Institutional capacity for climate change response : a new approach to climate politics / edited by Theresa Scavenius and Steve Rayner.
Description: Abingdon, Oxon ; New York, NY : Routledge, 2018. | Series: Earthscan science in society | Includes bibliographical references and index.
Identifiers: LCCN 2017040178 (print) | LCCN 2017052582 (ebook) | ISBN 9781315651354 (eBook) | ISBN 9781138120983 (hardback)
Subjects: LCSH: Climatic changes–Political aspects.
Classification: LCC QC903 (ebook) | LCC QC903 .I4785 2018 (print) | DDC 363.738/74561–dc23
LC record available at https://lccn.loc.gov/2017040178

ISBN: 978-1-138-12098-3 (hbk)
ISBN: 978-0-367-35835-8 (pbk)

Typeset in Times New Roman
by Wearset Ltd, Boldon, Tyne and Wear

Contents

Illustrations

Figures

Tables

Boxes

Contributors

Neil Adger is Professor of Human Geography at the University of Exeter, UK. His research spans the social dynamics of environmental change including political economy, demographic and resilience dimensions. He is a Highly Cited Researcher in the Social Sciences. He has served as a senior author to reports of the Intergovernmental Panel on Climate Change, on the Millennium Ecosystem Assessment, The Lancet Commission on Climate Change and Health, and the UK's Climate Change Risk Assessment. He is currently working on social dynamics of well-being, health, migration and climate change in the UK, India, Bangladesh, Ghana and Australia.

Jessica Nihlén Fahlquist is a senior lecturer in biomedical ethics at the Centre for Research Ethics and Bioethics at Uppsala University. Her research focuses on ethical aspects of risk in the context of technology, public health and environmental issues, and she has a particular interest in notions of moral responsibility. She has published articles on public health ethics, the philosophy of risk, the ethics of technology and environmental ethics. Jessica received her PhD in Philosophy at the Division of Philosophy at the Royal Institute of Technology in Stockholm in 2008. During the years 2007–2011 and 2012–2015, she worked as a postdoctoral researcher at the Philosophy department at Delft University of Technology.

Peter Healey is a sociologist whose main interest is the governance of science and technology, particularly in an international context. He has recently been working on climate geoengineering and on the issues raised by public participation in scientific practice.

Jens Villiam Hoff is a Professor at the Department of Political Science, University of Copenhagen. His research covers areas such as sustainability and climate change governance, citizen and user participation, and the use of digital media in politics. He is currently engaged in research projects on 'Developing and Implementing the Green GDP' and 'Collective Movements and Pathways to a Sustainable Society'. His most recent publications in English include 'The Green "Heavyweights": the Climate Policies of the Nordic Countries' (in P. Nedergaard & A. Wivel (Eds.), *Handbook of*

Scandinavian Politics, Routledge 2017) and J. Hoff & Q. Gausset (Eds.) *Community Governance and Citizen-Driven Initiatives in Climate Change Mitigation*, Routledge 2015.

Riadadh Hossain is an international development practitioner with a focus on climate change adaptation, urban resilience, climate policy, climate finance and governance. He is currently a Programme Coordinator at the International Centre for Climate Change and Development (ICCCAD) based in Dhaka, Bangladesh where he is overlooking a number of projects on climate finance, loss and damage, urban resilience, etc. in varying capacities. He is also responsible for guiding Monitoring, Evaluation and Learning (MEL) at ICCCAD. Riadadh has previously supported the Cities Development Initiative for Asia (CDIA) in their capacity development and outreach work and was responsible for the development of climate financing profiles for cities in Asia, as well as the design and implementation of an internal know-how learning series regarding urban resilience and climate change.

Saleemul Huq has been the Director of the International Centre for Climate Change & Development (ICCCAD) since 2009 and intends to support the growing capacity of Bangladesh stakeholders, while enabling people and organisations from outside to benefit from training in Bangladesh. ICCCAD runs regular short courses as well as an MSc programme in Climate Change and Development. Dr Huq is also a Senior Fellow at the International Institute for Environment & Development (IIED) in the UK, where he is involved in building negotiating capacity and supporting the engagement of the least developed countries (LDCs) in UNFCCC, including through holding negotiator training workshops for LDCs, policy, as well as research into vulnerability and adaptation to climate change in the least developed countries. Dr Huq has published numerous articles in scientific and popular journals, was a lead author of the chapter on Adaptation and Sustainable Development in the third assessment report of the Intergovernmental Panel on Climate Change (IPCC), and was one of the coordinating lead authors of 'Inter-relationships between adaptation and mitigation' in the IPCC's Fourth Assessment Report (2007).

Malene Rudolf Lindberg gained an MSc in Sociology (University of Copenhagen, 2015). She has worked as research assistant on the research project 'Mapping Value–Action Gap' at the Department of Political Science, University of Copenhagen, and has contributed to articles on climate behaviour and governance. Currently, Malene is working as a research assistant at The National Centre of Applied Social Science. Lindberg has studied in Copenhagen and at the Department of Social Sciences at Humboldt University of Berlin.

Irene Lorenzoni is Senior Lecturer at the University of East Anglia, UK. Her research focuses on individual, social and institutional engagement with environmental change; including perceptions of and responses to climate change and flooding; understandings of risk; and the evolution of climate policy. Irene's work is theoretically and methodologically interdisciplinary. Her work is highly cited; she was a contributing author to the IPCC 4th Assessment Report. Her current work focuses on responses to multiple hazards, climate change communication, comprehensibility of climate graphics, and diffusion of climate policy.

Conor Murphy is Senior Lecturer in Geography at Maynooth University and PI with the Irish Climate Analysis and Research UnitS (ICARUS). His research interests span the physical and human dimensions of climate change, from the detection and attribution of climate change from observations to understanding the social dynamics of adaptation to extreme events. At Maynooth University Conor also directs the MSc in Climate Change. He has worked with communities in Ireland, Europe and Africa on climate change adaptation. He sits on the Irish National Adaptation Committee and is on the editorial board of the Journal of Extreme Events. Conor is also review Editor for the Urban Climate Change Research Network.

Naznin Nasir is ICCCAD's migration programme coordinator. She joined ICCCAD in June 2016 and has since been involved with several programmes of the Centre such as the Climate Governance, Climate Finance, and Loss and Damage Programme. Naznin previously worked as a Research Associate with the Centre for Climate Change and Environmental Research at BRAC University, Dhaka. She received her Master's from the University of Western Ontario, Canada in Environment and Sustainability. She did another Master's in International Relations at the University of Dhaka, Bangladesh. Naznin has also worked as a broadcast journalist for the past 10 years.

Steve Rayner is James Martin Professor of Science and Civilisation at Oxford University's School of Anthropology and Museum Ethnography where he directs the Institute for Science, Innovation and Society. He previously held senior research positions in two US National Laboratories and has taught at leading US universities, including Cornell, Virginia Tech and Columbia. He has served on various US, UK and international bodies addressing science, technology and the environment, including Britain's Royal Commission on Environmental Pollution, the Intergovernmental Panel on Climate Change and the Royal Society's Working Group on Climate Geoengineering. Until 2008, he also directed the national Science in Society Research Programme of the UK Economic and Social Research Council.

Alexander Ruser is currently a Senior Researcher at Zeppelin University Friedrichshafen and Postdoctoral Fellow at the OIS Institute Vienna. In 2016 Alexander was deputy professor for cultural theory and analysis at Zeppelin University, Friedrichshafen. He holds a PhD in Sociology from the Max-Weber Institute of Sociology at Heidelberg University and was a Dahrendorf Postdoctoral Fellow at the Hertie School of Governance in Berlin and at the London School of Economics and Political Sciences and a Visiting Fellow at

Punjab University, Chandigarh, India. Alexander's research focuses on exper-tise and decision-making in climate, economic and social politics and he has published in peer-reviewed journals such as *Global Policy, Current Sociology, Innovation* and *Journal of Civil Society* and is an active member of an international research network on social philosophy of science coordinated by the Russian Academy of Science.

Theresa Scavenius is associate professor in the Department of Planning, University of Aalborg Copenhagen, Denmark. She has published widely on climate politics, global justice, and the relationship between facts and norms. Her recent publications include 'Fact-Sensitive Political Theory', published in *CRISPP* (2017), 'The Issue of No Moral Agency in Climate Action', pub-lished in *Journal of Agricultural & Environmental Ethics* (2017), and 'The Tragedy of the Few', published in *Res Publica* (2016). She co-edited the book on *Compromise and Disagreement in Contemporary Political Theory* (Routledge New York, 2018).

Nico Stehr is Karl Mannheim Professor of Cultural Studies at the Zeppelin University, Friedrichshafen, Germany. He is a Fellow of the Royal Society (Canada) and the European Academy of Sciences and Arts. His research interests centre on the transformation of modern societies into knowledge societies. He is one of the authors of the *Hartwell Paper* on climate policy. His books have been translated into Chinese, Russian, Czech, Slovenian, Jap-anese, Hungarian and German. Among his recent book publications (in English) are: *Experts: The Knowledge and Power of Expertise* (with Reiner Grundmann, Routledge 2011); *The Power of Scientific Knowledge* (with Reiner Grundmann, Cambridge University Press, 2012); *Information, Power, and Democracy. Is Liberty a Daughter of Knowledge?* (Cambridge Univer-sity Press, 2016), *Understanding Inequality. Social Costs and Benefits* (with Amanda Machin, Springer, 2016); *Knowledge. Is Knowledge Power?* (with Marion Adolf, Routledge. 2017) and *Climate & Society* (with Amanda Machin and Hans von Storch, World Scientific, 2017).

Tara Quinn, Dr, is an Associate Research Fellow in Geography at the Univer-sity of Exeter, UK. Her research interests centre on experience of environ-mental change and include a focus on demographic change, sense of place and perceptions of risk.

Introduction

Theresa Scavenius and Steve Rayner

This book seeks to stimulate wider discussion of how to locate climate politics in the current landscape of social and political institutions. In a period of rapid climate change and shifting patterns of governance both nationally and internationally, it is essential to understand and address both how different institutions at various levels react to climate change, and how the characteristics of institutions both enable and constrain their ability to respond effectively to it.

Much scholarly attention has been paid to the urgent need for immediate implementation of mitigation and adaption climate policies at regional and global levels. There are repeated calls for new policy strategies and innovative, proactive solutions to overcome the current gap between rhetoric and action with regard to climate change policy. However, many of these policies neglect or underestimate the importance of understanding the capacity of institutions to recognise, respond and act upon climate change. The established paradigm is deeply rooted in methodological individualism, leading to two kinds of policy solutions that dominate current political debates, responses and practices – market-based emissions trading and market-based voluntarism (Ronit, 2012). While emissions trading was favoured for its supposed efficiencies in reducing greenhouse gas emissions, and worked well as a domestic instrument to reduce sulphur emissions in the USA, the anticipated global market in carbon and other greenhouse forcing agents shows no sign of becoming a reality more than two decades after the ratification of the Kyoto Protocol. The second set of market-based policy solutions, aimed at greening the production of goods and services by voluntary certification and offset schemes, is an example of a policy solution that transfers the moral responsibility to buy greener products to individual consumers. If the consumers are not willing to pay a higher price for the greener product, nothing is accomplished.

In contrast with these approaches, the institutional capacity approach adopted in this book seeks to understand how, why and when diverse institutions, rather than individuals, respond to threats and challenges such as climate change. It is a theoretical approach concerned with institutional opportunities for, and hindrances to, political action. One might develop any number of good ideas to better manage the multiple challenges presented by climate change, but if political, economic and social institutions are incapable of implementing them in an effective and legitimate manner, then they will have no impact.

The contributors to this book approach climate action and politics as a poly-centric, multi-level governance challenge. The failure of the Kyoto architecture has given rise to a growing recognition that an effective human response to climate variability and change cannot simply be driven by national governments agreeing and implementing international treaties. Indeed, numerous studies have illustrated the expanding role of sub-state and non-governmental actors in developing climate change responses (e.g. Rabe, 2004; Bulkeley & Betshill, 2005; Acuto, 2013). The past two decades have shown that it will require the engagement of multiple agents (citizens, municipalities, states, international institutions and organisations) at multiple institutional levels (neighbourhood, municipal, national, regional and international).

Calls for a multi-level 'polycentric' approach were made as long ago as the late 1980s (Gerlach & Rayner, 1988), and arguments for a specific focus on institutions date from the 1990s (Rayner, 1993; O'Riordan et al., 1998). However, it was not until the collapse of the Kyoto architecture at Copenhagen in 2009 and Nobel Prize Laureate Elinor Ostrom's (2012) call for a multi-level and polycentric approach to climate change and sustainability that scholars and activists began to take real notice of the roles and capabilities of a wide range of institutions, at multiple levels, in the effort to counteract climate change.

The study of institutions is central to understanding the organisation and functioning of all societies (O'Riordan et al., 1998). The meaning of institutions contains several highly interrelated concepts, such as norm regulation, cognitive structures, and facilitation of identity and meaning. The institutional approach is inherently interdisciplinary and combines empirical knowledge about people's behaviour with insights from theoretical models of agency. Each scholarly discipline focuses on particular nuances. In political science and law, institutions are often thought of as formal organisations, frequently associated with the state, or legal obligations, such as contracts and treaties between individuals or states. On the other hand, sociologists and anthropologists generally regard almost any persistent pattern or non-randomness of social behaviour as an institution (Thompson, 2008). As argued by DiMaggio and Powell (1991), institutional analysis requires a multidimensional theoretical approach.

It is important to distinguish 'economic' from 'sociological' and 'historical' forms of institutional theory. Economic institutional theory ultimately remains wedded to the single agent's behavioural choices and the instrumental rationality behind these choices. By contrast, sociological institutionalism focuses on the institutionalised structure which surrounds the acts of the agents and historical institutionalism on the historical continuity and the path dependencies to which institutions are thought to be subjected (Hall & Taylor, 1996).

A thoroughgoing institutional approach differs from rational choice theories that are derived from assumptions about individual behaviour, by rejecting both an overwhelmingly rationalistic understanding of agency and an exclusively economic approach to politics (Rayner, Lach, & Ingram, 2005). In contrast to rational-economic premises, the institutional perspective takes a non-reductionist approach to agency, i.e. that there is no single (rational) strategy that can explain people's behaviour (List & Pettit, 2011). Non-reductionism is a methodological strategy where different approaches to agency, causality and rationality are allowed to co-exist at various levels in agentic and institutional contexts. The assumption is that agents develop different behavioural choices depending on specific institutional circumstances (Le Grand, 2003) or, even more radically, that agency is inherently relational rather than individual (Thompson, 2008). The focus on institutions is an analytic strategy that emphasises how organisational and social arrangements both shape and are shaped by policies and practices, leading to different, and sometimes unexpected, outcomes.

For example, standard neo-classical economic or rational choice theory would predict that blood donation would increase where donors were offered an economic incentive to give blood. However, a classic study by Richard Titmuss (1970) found that people's inclination to give blood decreases when they are offered money for the blood. Blood donors appeared to frame their actions within the institution of gift giving rather than commercial transactions. Although more recent studies suggest that at least some donors do respond to incentives (e.g. Lacetera, Macis, & Slonim, 2013), these incentives still seem to be more successful when framed as reciprocal gifts rather than cash payments. The example of blood donation illustrates the importance of understanding quite subtle differences in the institutional framing of actions and objectives.

Climate change and the design of climate change responses pose similar challenges of understanding the institutional context and the subtleties of institutional design. Should climate action be considered an economic transaction where we are expected to get paid for all the sacrifices we accept, or rather as a moral concern where economic costs are less relevant? The answer is possibly both, depending on the institutional context in which the question is being posed. To whom is the question being addressed? We might expect different answers from members of close-knit communities, where mutual aid and ideas of the common good prevail, than the answers we would anticipate from dispersed individuals with much looser ties of friendship and kinship (Douglas, 1986). In either case, successful intervention depends on matching the assumption about motivation to the particular context. Getting this wrong is likely to render the intervention ineffective at best and, at worst, an offensive provocation. This is clearly illustrated in US climate politics, which have descended to a Manichean standoff between so-called sceptics, who reject the very idea of anthropogenic climate change because of what they see as the opportunity it presents for meddling in free markets and limitations on personal freedom, and climate activists, who view such rejection of the threat of climate change as an offence to rationality perpetrated by commercial interests tied to the fossil fuel industry. The institutional approach emphasises that different institutional commitments lead to very different orientations that have to be accommodated in pluralistic policy responses. Successful climate policy design works with the grain of social and political arrangements, rather than struggling against it (Kaul & Mendoza, 2003).

A Faustian bargain?

Human behaviour, industrial production and lifestyle choices affect the underlying dynamics of the planet's climate. Its most immediate threat is to vulnerable people and species in marginal environments, but left unchecked, it may irreversibly change the natural conditions under which modern civilisation is built. To many, the future looks bleak if the current generation proves to be incapable of curbing climate change and its negative effects on the earth's natural resources.

For the past two decades, international efforts to avoid 'dangerous anthropogenic interference with the climate system' (UNFCCC 1992 Art 2) have been conducted under the auspices of the UN Framework Convention on Climate Change. Implementation of the convention's goals was to be achieved through the 1997 Kyoto Protocol, which finally entered into force in 2005 when Russia signed the treaty.

The FCCC–Kyoto architecture was based on an earlier treaty to protect the atmosphere, not against climate change, but against depletion of the protective stratospheric ozone layer due to the emissions of chlorofluorocarbons (CFCs) and other ozone-depleting substances. The Montreal Protocol paved the way for banning the production of CFCs used in mundane applications such as aerosol cans, refrigeration and air conditioning equipment.

While the Montreal Protocol has been successful in reducing CFCs, the Kyoto Protocol conspicuously failed to reduce greenhouse gasses (Prins & Rayner, 2007). CFC levels peaked in 1994; since then, these gases have been gradually decreasing. It is estimated that the Antarctic ozone layer will be completely recovered by 2050 or 2068 (Newman, Nash, Kawa, Montzka, & Schauffler, 2006). The opposite applies in the case of greenhouse gases (GHGs), whose emissions have grown steadily in recent decades. According to the Intergovernmental Panel on Climate Change (IPCC, 2014), global annual GHG emissions of around 49 gigatonnes (GT) (2010) come directly from energy supply (47%), industry (30%), transport (11%) and agriculture, land use and deforestation (24%).

The world's leaders very publicly failed to agree on a legally binding global agreement at the fifteenth session of the Conference of the Parties (COP15) to the UNFCCC, which took place in December 2009 in Copenhagen. COP15 has become a reference point in the international academic and policy communities as confirming the failure of the Kyoto approach to climate policy.

The failures of global climate politics have been interpreted in some quarters as a failure of democratic governance (Hoff & Strobel, 2013), and, more specifically, the ineffectiveness of democratic decision-making processes: 'Democracy is inconvenient and identified as the culprit holding back action on climate change' (Stehr, 2013). But to take such a view is to equate effective measures to protect the climate with the conclusion of treaties. A democratically achieved treaty would still fail to protect the world from climate change if the institutional frameworks for its implementation were flawed. It is also possible to conceive of institutional actions leading to positive climate outcomes without the need for a single universal treaty, as suggested by Gerlach and Rayner 30 years ago.

The Paris Agreement of 2015 signals a move towards polycentricity by including, for instance, 'non-market based' policy designs. However, to be effective, the Paris approach needs to be underpinned by wider and more systematic understanding of the contributions that institutions of all sorts and democratic decision-making processes can have (Nasir, Hossain, & Huq, Chapter 8, this volume).

Nevertheless, some commentators argue that democratic decision-making processes take too long and are too complicated to be effective. They variously argue in favour of a rapid technological fix, such as the injection of sulphate aerosols into the atmosphere to reflect some of the sun's energy away from earth (Crutzen, 2006), or for a programme of forced technological change and consumption controls imposed by a 'good authoritarian' leader (Beeson, 2010, p. 289). The suggestion is that '[w]e need an authoritarian form of government in order to implement the scientific consensus on greenhouse gas emissions', and 'humanity will have to trade its liberty to live as it wishes in favor of a system where survival is paramount' (Shearman & Smith, 2007, p. 4).

However, such a rejection of democracy is premature from a social scientific point of view. Prins and Rayner (2007) argue that the failures of the Kyoto Protocol cannot be explained by failures of democratic commitment. Instead, the Protocol failed because its institutional design was not suitable for managing GHG emissions globally. It is even arguable that one of the problems of climate politics might be not too much but too little democracy, as the current political-economic order favours entrenched interests in fossil fuels. If that is the case, one may wonder why anyone would blame democracy for the failures of global climate action. One place to look for a clarification of why democracy has become a scapegoat might be to question the dominant framing of the climate change problem.

A recent scientometric study of the IPCC Third Assessment report demonstrates a scientific and economic bias in climate research (Bjurström & Polk, 2011). Research in climate change is dominated by the earth sciences, economics and, to some extent, by moral philosophy. The earth scientists' models predict climate changes, the economists calculate the costs of various climate policies, and the moral philosophers analyse the moral blame and responsibilities of polluters. These three groups of scholars share an approach to climate change that is individualistic and emissions-oriented. The focus on individuals' behaviour and the costs and effects of emissions establishes a scientific focus that generally elides the social and political context of human behaviour.

The problem is not that these disciplines focus on their specific research interests, but that many of the practitioners do not address the relevance of political and institutional analyses of climate change. Some have called this neglect a 'Faustian bargain', where climate scholars and policy-makers do not use the full range of social science theories of social agency and institutional frameworks, but narrowly focus on a few ideas about the economic valuing of greenhouse gas emissions (Sagoff, 2008). Climate research and policy frequently succumbs to 'climate reductionism' that reduces the human complexity of the issue to a simple 'end-of-the-pipe' problem of stopping pollution.

Against this reductionism, Stehr and Von Storch argue that

> [c]oncentrating climate policy on the reduction of greenhouse gases serves
> no purpose if it leads at the same time to preventing taking precautions in
> dealing with present dangers and their possible future amplifications. Such a
> one-sided perspective and climate protection policy will neither protect the
> climate from society in the coming decades, nor society from the climate.
>
> (Stehr & Von Storch, 2009, p. 57)

Two examples of the present dangers referred to by Stehr and Von Storch
concern the heat-related deaths in the blistering summer of 2003 in Europe and
the forest fires in Russia 2007. The popular explanations of these events at the
time were global warming. Little, if anything, was said about municipalities,
regions, or countries that failed to take the necessary precautions before these
catastrophic events occurred. The problem with neglecting such intermediate
factors as taking precautions is that the social scientific analysis of institutional
and political contexts and preconditions is overlooked. The social sciences have
distinct explanatory models that are not reducible to a linear causal relation
between a natural cause, such as global warming, and the consequences of heat
stroke in Paris or forest fires in Russia. But, as Stehr (2013, p. 58) argues, 'by
focusing on the effects or goals of political action rather than its conditions, the
contentious issue of climate change is reduced from a socio-political to a technical issue'. The result is the depoliticisation of climate change and a sceptical attitude towards democratic climate governance. The failure of democracy is seen
as a cause of the failure of climate action.

Another example of reductionism is the claim that climate change affects poor
people more dramatically than rich people. It is recognised both by scientists (see
for example Riebeek, 2010) and public media (see for example Worland, 2016)
that the poorest people on earth will also be hardest hit by global warming and
rising sea levels, which is predicted to be a consequence of climate change.
Again, this explanation obscures the broader set of social and institutional mediating factors. The fact that poor people are at higher risk of climate change is
taken for granted. However, the reason why poor people are at higher risk is not
because climate change and the ensuing rising sea levels hit the Maldives harder
than the Netherlands. The reason is that the Netherlands is politically and institutionally better prepared to adapt to the changes by securing areas at risk.

So why focus on institutions?

An institutional perspective is required to understand the differences between the
Maldives and the Netherlands. Furthermore, an institutional approach suggests
that climate action failures have their roots in 'the systematic problem-solving
gap' which challenges many countries today (Scharpf, 2006, p. 856). In particular, market-based policies that stimulate global market competition in public
institutions and post-national rights regimes have, as Plant has demonstrated,

eroded the powers formerly available to state institutions to correct political and economic imbalances resulting from the operations of the market within their own borders. Indeed, many of the rules agreed to under free trade agreements effectively prevent states unilaterally adopting progressive social, economic or environmental legislation.

(Plant, 2010, p. 14)

In the last decade or so, research has advanced our understanding of institutional designs as frameworks for agency (Ostrom, 2005) and institutions as collective agents (List & Pettit, 2011). The common denominator for this new interest in institutions is the meso-level explanatory model of social and political agency (Rayner & Malone, 1998, vol. 4, p. 46). Institutionalism draws attention to a level of political organisation, governance and agency that cannot be reduced to micro approaches featuring political agency as rational single policy-makers; nor can it be reduced to macro approaches that give primacy to societal forces of economic competition, technological innovation and demographic change (Olsen, 2007). Institutionalism maintains that formal and informal institutions constitute a sui generis collective level of agency (Searle, 2010), which means that institutions are not epiphenomena of individual preferences or underlying production circumstances. In contrast, they supervene on the individual agents.

Institutional analysis offers understanding of the mediating factors that either provide or confine the negative consequences of climate change. A multidisciplinary and multi-level approach can address the current social and political challenges of curbing climate change and the deteriorating conditions of the earth's natural resources and sentient life. Political and institutional analyses contribute to the debate by drawing attention to what consequences climate change and climate politics may have for citizens' sovereignty, political institutions and the future of democracy. When institutional analysis is added to the research mix, climate change is not merely a matter for scientists studying the degree of global warming or economists calculating the most efficient mitigation policies. It is also a major interdisciplinary theoretical topic that should challenge our intuitively correct conceptualisations of politics as they are represented in contemporary political science and moral philosophy.

The practical gain from moving beyond the individualistic and emissions orientation to a multi-level, multi-faceted paradigm is that a broader range of policy solutions is revealed.

Institutional capacity

In recent years, there has been an increasing amount of literature on issues related to the concept of institutional capacity (Adger & Jordan, 2009; Pelling, 2010; Rayner & Caine 2015). These contributions take a less restrictive approach to agency and rationality than has been the case in the classic accounts of economic and sociological institutionalism. Whereas the distinction between economic and sociological institutionalism is defined by disagreements about rationality, causality and agency, the new interest in institutional capacity, capabilities and resilience appreciates various types of methodological ideas regarding rationality, causality and agency. Climate change scholarship is paying increasing attention to the issue of how to understand what it is that institutions are capable of facilitating, governing and maintaining. Depending on their design and construction, these capabilities enhance or impede various types of rationality and choices.

Complex societies have reached a high level of institutional capacity. For example, political institutions currently pool resources and manage the collective challenges of security, law and government. A highly sophisticated division of labour currently in place in public institutions enables complex decision-making, problem solving and management.

Theorisation of institutional capacity since the 1980s, when climate change emerged as a policy issue, has been developed in several disciplines. The first wave of capacity studies originated in anthropology and focused on institutional culture; the second was centred on political geographers who understood capacity as characterised by system properties. Building upon these pioneer works, the third wave of capacity studies, exemplified here, draws attention to political scientific aspects, such as questions of governance, democratic legitimacy, decision theory, and the institutionalisation of group agency.

Although it had strong roots in the cultural theory of institutions proposed by Mary Douglas (1986), the anthropological perspective was from the outset strongly interdisciplinary. Contributors to the special edition of *Evaluation Review* on Managing the Global Commons (Rayner, 1991) included two anthropologists, two international lawyers, two geographers, a political scientist and an STS scholar. Cross-national and comparative institutional issues were central concerns. A special edition of *Global Environmental Change* (Rayner, 1993) focused on specially commissioned empirical studies of four major countries and the European Community to assess how cultural, legal and political institutional factors could impede or enhance their capacity to adopt and implement initiatives to grapple with climate change. This approach was carried forward in subsequent work, most prominently presented in the four-volume book *Human Choice and Climate Change* (Rayner & Malone, 1998) and later the *Hartwell Approach to Climate Policy* (Rayner & Caine, 2015), representing a broadly institutionalist perspective.

Complementary to the broad cultural focus of the anthropological approach, many political geographers, such as Tompkins and Adger (2005) and Bulkeley (2013), identify criteria and indicators of how to measure the level of capacity of various systems, be they biological or institutional. The political geography of capacity studies was developed, in particular, within the framework of adaptation studies in the 2007 IPCC Fourth Assessment Report (FAR) by the Intergovernmental Panel on Climate Change (IPCC). In this report, adaptation studies are contrasted with mitigation studies, which focus on the economic costs of investments in technological equipment. They address how effective agents, institutions and cultures respond to climate changes. Adger et al. identify in the FAR the institutional ability to respond as the key point in the definition of adaptive capacity: 'Adaptive capacity is the ability or potential of a system to respond successfully to climate variability and change, and includes adjustments in both behaviour and in resources and technologies' (Adger, Agrawala, Mirza, Conde, O'Brien, Pulhin, et al. 2007, p. 727). They also define adaptive capacity as a system; the system understanding of capacity resembles approaches drawn from resilience, vulnerability, and sustainability studies of natural phenomena. Water, for example, is a system vulnerable to pollution and the washing out of nutrients caused by industrialised farming. Similarly, countries' adaptive capacities are considered as a system vulnerable to the lack of technological competences and financial resources. One system-centred proposal suggests two criteria to distinguish between low and high capacity: 'willingness' and the 'ability' of society to change. Willingness and ability are connected to 'the availability and penetration of new technology' (Tompkins & Adger, 2005, p. 564). The system-centred idea is that an increase in willingness and ability will increase the capacity to respond to climate change.

Overview of the book

This book brings together a broad range of social scientific approaches and methods to explore the content, scope and conceptualisation of the role institutions play in solving climate change. The multidisciplinary approach includes contributions from scholars of anthropology, political science, geography and environmental studies. Some contributors to this collection are primarily empirical social scientists, focusing on explicitly political institutions, while others are primarily social and political theorists, drawing attention to a broader range of social arrangements. In selecting the chapters in this book, the editors have sought to reflect this spectrum of social science thought in exploring the concept of institutional capacity to respond to climate change.

The contributions represent a cross-section of work in the field and address institutional aspects of democratic and political governance relevant to the development of effective climate change programmes. Early versions were presented at the Copenhagen Climate Workshop Series, sponsored by the Villum Foundation at the Department of Political Science, University of Copenhagen in 2013–2014. The focus of the discussion at the workshops was on the need and potential for a specifically institutional focus for understanding contemporary climate politics and its failures.

The book is organised into three parts: the first explores the institutional capacities of democracies, the second examines institutional roles in achieving collective behavioural change, and the third addresses some of the institutional challenges of developing climate response strategies.

In Part I *Institutional Capacity and Democracy*, the first chapter by Stehr and Ruser discusses the limits and benefits of democratic governance by pointing towards the complex and delicate interplay between (expert) knowledge and decision-making; thus investigating the reasons and the possible consequences of an uneasiness with democracies in the era of anthropogenic climate change. In Chapter 2, Scavenius aims to demonstrate that the critiques of contemporary democratic climate governance converge in their neglect of the importance of maintaining the democratic capacity of political institutions, such as parliaments and other formal institutions of government. In Chapter 3, Nihlén Fahlquist explores the principles of institutional responsibility. She raises the question of to what extent individual and institutional agents are morally responsible for climate change, in different ways and for different reasons.

Part II *Response Capacity and Behavioural Change* contains three chapters that touch on the difficulties of building response capacity through behavioural change. In Chapter 4, Scavenius and Lindberg argue, on the basis of a Danish survey study, that many climate actions, both successful and unsuccessful, remain underexplained at the level of individual agency. Presenting a study of 360 households across Ireland and England, Adger and co-authors examine in Chapter 5 how perceptions of government action held by citizens influence the citizens' decision to take actions themselves. In Chapter 6, Hoff analyses the level of citizenship involvement in recent climate change adaptation strategies in New York.

The book closes with Part III *Institutional Capacity in Society* in which two chapters argue that the institutional capacity approach fosters new perspectives on concrete institutions and policy areas. In Chapter 7, Rayner and Healey take a broad approach to institutions to describe the governance challenges for developing geoengineering proposals. They conclude that the debate on geoengineering presents a valuable opportunity to stimulate a wider institutional discourse about key issues such as the relationships between humans and nature, and technology and society. In Chapter 8, Nasir and co-authors explore the potential role that universities could play in building the capacity of the next generation, as well as today's policy-makers and implementers, in dealing with climate change.

These chapters and their different perspectives illustrate just some of the ways in which an explicitly institutional perspective can enrich our understanding of the challenges and opportunities facing humanity in understanding and managing the sources and consequences of climate change and, in so doing, contribute to the opening up of climate change policy to multi-level, polycentric responses made possible by the change in policy architecture signalled by the Paris Agreement.

References

Acuto, M. (2013). *Global cities, governance and diplomacy: The urban link*. Routledge.
Adger, N., & Jordan, A. (2009). *Governing Sustainability*. Cambridge: Cambridge University Press.
Adger, W. N., Agrawala, S., Mirza, M. M. Q., Conde, C., O'Brien, K., Pulhin, J., Pulwarty, R., Smit, B., & Takahashi, K. (2007). Assessment of adaptation practices, options, constraints and capacity. In M. Parry et al. (Eds.), *Climate change 2007: impacts, adaptation and vulnerability*. Cambridge University Press.

Beeson, M. (2010). The coming of environmental authoritarianism. *Environmental politics, 19*(2), 276–294.

Bjurström, A., & Polk, M. (2011). Physical and economic bias in climate change research: A scientometric study of IPCC Third Assessment Report. *Climatic Change, 108*(1): 1–22.

Bulkeley, H. (2013). *Cities and Climate Change.* New York: Routledge.

Bulkeley, H., & Betsill, M. M. (2005). *Cities and climate change: urban sustainability and global environmental governance* (Vol. 4). Psychology Press.

Crutzen, P. J. (2006). Albedo enhancement by stratospheric sulfur injections: a contribution to resolve a policy dilemma? *Climatic change, 77*(3), 211–220.

DiMaggio, P. J., & Powell, W. W. (Eds.) (1991). *The new institutionalism in organizational analysis.* University of Chicago Press.

Douglas, M. (1986). *How institutions think.* Syracuse University Press.

Gerlach, L., & Rayner, S. (1988). Culture and the common management of global risks. *Practicing Anthropology, 10*(3–4), 15–18.

Hall, P. A., & Taylor, R .C. (1996). Political science and the three new institutionalisms. *Political studies, 44*(5), 936–957.

Hoff, J. V., & Strobel, B. W. (2013). A Municipal Climate Revolution'? The Shaping of Municipal Climate Change Policies. *The Journal of Transdisciplinary Environmental Studies, 12*(1), 4–16.

Kaul, I., & Mendoza, R. U. (2003). Advancing the concept of public goods. *Providing global public goods: Managing globalization, 78*, 95–98.

Lacetera, N., Macis, M., & Slonim, R. (2013). Economic rewards to motivate blood donations. *Science, 340*(6135), 927–928.

Le Grand, J., (2003). *Motivation, agency, and public policy: of knights and knaves, pawns and queens.* Oxford University Press on Demand.

List, C., & Pettit, P. (2011). *Group agency: The possibility, design, and status of corporate agents.* Oxford University Press.

Newman, P. A., Nash, E. R., Kawa, S. R., Montzka, S. A., & Schauffler, S. M. (2006). When will the Antarctic ozone hole recover? *Geophysical Research Letters, 33*(12).

Olsen, J. P. (2007). Understanding institutions and logics of appropriateness: Introductory essay. Working Paper 13, Centre for European Studies, Oslo.

O'Riordan, T., Cooper, C. L., Jordan, A., Rayner, S., Richards, K. R., Runci, P., & Yoffe, S. (1998). Institutional frameworks for political action. In S. Rayner & E. L. Malone (Eds.), *Human choice and climate change,* Vol. 1. Battelle Press.

Ostrom, E. (2005). Doing institutional analysis digging deeper than markets and hierarchies. In C. Ménard & M. M. Shirley (Eds.), *Handbook of new institutional economics* (Vol. 9). Springer.

Ostrom, E. (2012). Nested externalities and polycentric institutions: must we wait for global solutions to climate change before taking actions at other scales? *Economic theory, 49*(2), 353–369.

Pelling, M. (2010). *Adaptation to climate change: from resilience to transformation.* Routledge.

Plant, R. (2010). *The neo-liberal state.* Oxford University Press on Demand.

Prins, G., & Rayner, S. (2007). Time to ditch Kyoto. *Nature, 449*(7165), 973–975.

Rabe, B. G. (2004). *Statehouse and greenhouse: The emerging politics of American climate change policy.* Brookings Institution Press.

Rayner, S. (Ed.) (1991). *Evaluation Review, 15*(1), special issue: Managing the global commons.

Rayner, S. (Ed.) (1993). *Global Environmental Change, 3*(1), special issue: National case studies of institutional capabilities to implement greenhouse gas reductions.

Rayner, S., & Caine, M. (Eds.) (2015). *The Hartwell Approach to Climate Policy.* Routledge.

Rayner, S., & Malone, E. L. (1998). *Human Choice and Climate Change*, Vols 1–4. Battelle Press.

Rayner, S., Lach, D., & Ingram, H. (2005). Weather forecasts are for wimps: why water resource managers do not use climate forecasts. *Climatic Change, 69*(2), 197–227.

Riebeek, H. (2010). Global Warming: How Will Global Warming Change Earth? NASA Earth Observatory. Link: https://earthobservatory.nasa.gov/Features/GlobalWarming/page6.php

Ronit, K. (Ed.) (2012). *Business and climate policy: The potentials and pitfalls of private voluntary programs.* United Nations University Press.

Sagoff, M. (2008). *The economy of the earth. Philosophy, Law, and the Environment.* Cambridge University Press.

Scharpf, F. (2006). The 'Joint-Decision Trap Revisited', *Journal of Common Market Studies, 44(*4), 845–864.

Searle, J. (2010). *Making the social world: The structure of human civilization.* Oxford University Press.

Shearman, D. J., & Smith, J. W. (2007). *The climate change challenge and the failure of democracy.* Greenwood Publishing Group.

Stehr, N. (Ed.) (2013). *Eduard Brückner – The sources and consequences of climate change and climate variability in historical times.* Springer Science & Business Media.

Stehr, N., & von Storch, H. (2009). Climate Protection. *Journal für Verbraucherschutz und Lebensmittelsicherheit, 4*(1), 56–60.

Thompson, M. (2008). *Organising and Disorganising. A Dynamic and Non-Linear Theory of Institutional Emergence and Its Implication.* Triarchy Press.

Titmuss, R. M. (1970). *The gift relationship. From human blood to social policy.* Allen and Unwin.

Tompkins, E. L., & Adger, W. N. (2005). Defining response capacity to enhance climate change policy. *Environmental Science & Policy, 8*(6), 562–571.

Worland, J. (2016, 5 February). How Climate Change Unfairly Burdens Poorer Countries. *Time Magazine.* Link: http://time.com/4209510/climate-change-poor-countries/

Part I

Institutional capacity and democracy

1 Climate change, governance and knowledge

Nico Stehr and Alexander Ruser

In the 1998 Jerry Bruckheimer blockbuster movie *Armageddon* humanity is faced with the most serious threat: an asteroid the size of Texas is on a collision course with the earth. The effects of such a collision would be nothing short of disastrous: the end of the world as we know it. In this situation people do the obvious: they turn to a hero (naturally: Bruce Willis) who eventually saves the planet.

If one believes in the IPCC reports and listens to the speakers at the numerous climate summits in the past two decades humanity is facing an equally serious crisis: the dangers and risks of anthropogenic climate change.[1]

Unlike the impact of an asteroid, the effects of climate change are not immediately catastrophic. Likewise, it is hard to attribute a particular disaster (say an extreme weather occurrence such as flood or a drought) to *climate* change. Nevertheless, anthropogenic climate change brings about *extraordinary dangers*. Another deviation from the Hollywood script is the notable (some might say deplorable) absence of a hero. Decisive policy action is still lacking and progress in international climate negotiations is slow, fragile and, sometimes, thwarted by short-term national interest.

Discussions about the governance of climate change in the coming decades increasingly appeal to the strong probability of exceptional dangers that modern societies are bound to face in the not too distant future, unless drastic and immediate action is taken to reduce greenhouse gas emissions around the world. What is of interest to us is that the appeal to extraordinary circumstances justifies two assertions, that of an inconvenient mind and that of inconvenient social institutions. These deficits in turn allow for and tolerate fundamental doubts about the efficacy of democratic governance to cope with extraordinary dangers.

Robert Stavins, director of Harvard's Environmental Economics Program and a co-author of the IPCC (Intergovernmental Panel on Climate Change) WG 3 report might well be quoted as a warrantor of this scepticism about the ability of political decision-makers to come up with timely and determined action:

> This bottom up demand which normally we always want to have and rely on in a representative democracy, is in my view unlikely to work in the case of climate change policy as it has for other environmental problems.... It's going to take enlightened leadership, leaders that take the lead.[2]

We can assume that Robert Stavins prefers the scripted reality of a Hollywood movie over the often frustrating complexity of political debate. In fact, when it comes to global climate politics, Bruce Willis is nowhere in sight. This leaves us with some important questions. Who could the enlightened leaders be? And, where would it lead us if we would surrender democratic rules in order to let 'enlightened leaders' take over?

In order to answer these questions we will advance our argument in four steps: first we'll investigate the relation between scientific knowledge and politics, for scientific experts are instrumental for the enlightenment of leaders, or could become such leaders themselves. We will then outline the historical roots of the inherent tension between the inclusiveness and quality of democratic decision-making to explain (third) the 'discreet charm' of expert rule today. Finally, we will point at the problems at risk to indicate the obvious faults of the assertion of an inconvenient democracy.

Scientific knowledge and politics

A suitable starting point for investigating the interplay of scientific knowledge and political decision-making is to focus on the evolution of the social sciences. When in the 19th and early 20th century the 'social sciences' emerged and differentiated themselves from the humanities, the vanishing point usually was the natural sciences (Ruser, 2016). August Comte, for instance envisioned a 'social physics' rather than sociology. The explicit aim was to subject 'social phenomena, like all others, to invariable natural laws (…) in other words, introducing into the study of social phenomena the same positive spirit which has regenerated each other branch of human speculation' (Comte, 1855, p. 455).

However, since then scholars have tended to emphasise theoretical and epistemological dissimilarities and differences between the social and the natural sciences (see Snow, 1959). Consequentially most social scientists today are reluctant to make statements about future economic, political and social conditions (which at one time were taken to be part and parcel of genuine social science discourse, see Elias, 1987).

Modern social scientists have good reasons for their reservations. From the perspective of philosophy of science, insightful contributions like Robert K. Merton's 'The self-fulfilling prophecy' (1948) have revealed that sociological description (of trends) bears the risk of prescribing future developments (Ruser, 2015a). The interplay between social scientific expertise and public deliberation and decision obviously evades simple conceptions of 'speaking truth to power' (Wildavsky, 1979).

However, we are confronted with a new situation. Anthropogenic climate change represents a global challenge that transgresses national and disciplinary boundaries alike. Since the 1970s most of the scientific discourse has been devoted to establishing the phenomenon. Now that the phenomenon has been established,[3] discussion, as the most recent Assessment Report of the IPCC (2014) demonstrates, focuses on ways of averting or reducing the impact of climate change. Still in the doghouse of the climate change discussion is the issue of adaption and, significantly, governing (or coping) with the future consequences of climate change. It is precisely because the governance of climate change is highly dependent on scientific expertise and has major social, political and economic implications that many people who 'have the facts' grow increasingly impatient with national and international politics, which fall short of the 'necessary' climate objectives. Is it time for scientists to become less reluctant? Should expert knowledge trump public sentiment to overcome cumbersome debates and allow the problem to be tackled?

As noted above, circumstances are 'extraordinary dangers'. Normally, in everyday life in modern society the layperson has but few encounters with, or searches for, novel knowledge claims that prompt the actor to critically reflect on the validity of the claim or question the authority of the expertise of the actor advancing the claim. It is of course not ruled out completely that conflicts emerge in such everyday contexts and searches for knowledge. But such conflicts tend to be extraordinary. Often, such conflicts are resolved by appealing to the normality of the case (for example, the overwhelming evidence) under dispute. In the case of climate change, and especially its likely consequences, we are confronted with entirely different assertions. Except for reference to singular historical events, for example war, there are no normal experiences to which the claims can appeal, as would be the case, say, in evidence-based medicine. Governing the consequences of climate change relates to a time scale and societal transformations that are beyond the imagination essential to cope with everyday life. This implies that the relation between scientific knowledge and power demands particular attention: specialised knowledge that advances claims about far-reaching (political) consequences and makes statements about a distant future directly affect the authority of contemporary (political and/or economic) elites. At least in democratic systems 'climate expertise' is inevitably political. Scientific knowledge claims can augment or undermine the authority of power elites and electorates alike. Climate knowledge is at the same time a *strategic resource* and a *serious threat* depending on one's position on the political spectrum.

The inherent tension between participation and expertise

Voices ready to sacrifice democratic rights and liberties in the face of urgent societal problems are nothing new. Democratic deliberation and universal suffrage have frequently been called into question, when pressing issues demanded quick and, above all, 'the right' decisions.[4] The underlying assumption is that in a situation of crisis one cannot afford to make uninformed decisions. However, as John Adams noted: 'public information cannot keep pace with the facts' (Adams cited in Powell, 1951, p. 531). In different historical context, decades ago, the economist Friedrich Hayek, ([1960] 2006, p. 25) pointed to the paradoxical development with respect to the relation of knowledge and liberty that tends to follow scientific advances; scientists, as well as others, labour under the illusion that the volume of our ignorance is steadily diminishing with the advance of knowledge. This tends to strengthen that view that we should 'aim at more *deliberate and comprehensive control of all human activities*'. Hayek pessimistically adds, 'It is for this reason that those intoxicated by the advance of knowledge so often become the enemies of freedom.'

It is important to note that 'enemies of freedom' must not be hostile to civil liberties per se. Rather, it is the circumstances that call for bold and decisive action, even if this means (temporarily) suspending basic democratic rights. This line of argumentation derives from Isaiah Berlin's ([1958] 1969, p. 134) comment that if those who disagree were only more enlightened – that is, prepared to take on board the 'objective' framing of options – [those] individuals and groups would pursue the same course of action. As Berlin therefore stresses, such a state of affairs 'renders it easy for me to conceive of myself as coercing others for their own sake, in their, not my, interest. I am then claiming that I know what they truly need better than they know it themselves.' If people would only understand the causation and the consequences of climate change, they surely would make the inevitable choice to back the goals and policies of global climate politics. What could be wrong with an enlightened elite making decisions, if they in fact do know better?

As said above, the conviction that knowledgeable people are better equipped to ensure proper government and should, therefore, be allowed to act on behalf of the less educated is not new. In fact, even one of the most prominent figureheads of liberal thinking, John Stuart Mill shared this point of view. In his thoughts on *Representative Government* (1861) he directly links suffrage to the (lack of) intelligence and education, claiming that well-educated voters are better suited to act on behalf of the uneducated mass:

> The opinions and wishes of the poorest and rudest class of labourers may be very useful as one influence among others on the minds of the voters, as well as on those of the Legislature; and yet it might be highly mischievous to give them the preponderant influence by admitting them, in their present state of morals and intelligence, to the full exercise of the suffrage.
>
> (Mill, 1861, p. 129)

It is important to stress that the notion of an 'inconvenient democracy' is linked to disenchantment with global climate politics. It is equally important to emphasise that this line of argumentation has a much longer history.

Nevertheless, the urgency and the magnitude of the problem understandably worry climate scientists all over the world. The renowned climatologist Hans-Joachim Schellnhuber (2011)[5] for instance, is concerned with the progress of global climate politics because 'my own experience and everyday knowledge illustrate that comfort and ignorance are the biggest flaws of human character. This is a potentially deadly mix.' In other words, there are considerable obstacles which prevent 'the people' from taking on board the 'objective' framing of options as longed for by Berlin.

So far, voices advocating more technocratic decision-making on climate politics have been only marginal. A notable exception is Evelyn Fox Keller (2010), who made the strong case for an immediately effective political role for climate science, given the seriousness of the problem of global warming:

> there is no escaping our dependence on experts; we have no choice but to call on those (in this case, our climate scientists) who have the necessary expertise.... Furthermore, for the particular task of getting beyond our current impasse, I also suggest that climate scientists may be the only ones in a position to take the lead. Finally, given the tacit contract between scientists and the state which supports them on the other, I will also argue that climate scientists are not only in a position to take the lead, but also that they are obliged to do so.
>
> (Keller, 2010, 2011)[6]

From discomfort with democracy to an inconvenient democracy?

To say that technocracy isn't widely embraced doesn't imply that there is no criticism of how democratic systems have dealt with the challenge of anthropogenic climate change so far. Moderate critics point towards the 'tendency (...) to present (...) preferred "facts" and associated policy as true and obvious' (Lahsen, 2005, p. 141), thus questioning the compatibility of climate science research and democratic decision-making. Likewise, uncritical appraisal of democracy as guaranteeing political participation is rejected as a 'tendency to celebrate "civil society" without attending to the role of power inequalities' (Lahsen, 2005, p. 159). The capacity of democracies to address climate change is, from the perspective of these moderate critics, limited by two aspects: (1) the difficulty in feeding scientific expertise into the political process and (2) the concealment of power inequalities.

More fundamental critics deny democratic systems' ability to cope with global climate change altogether: according to Shearman and Smith (2007) there is 'much evidence to suggest that liberal democracy – the meshing of liberalism and democracy – is the core ideology responsible for the environmental crisis' (Shearman & Smith, 2007, p. 12). Shearman and Smith argue that democratic institutions are by design incapable of dealing with crisis; that is, to provide working solutions when quick and determined decision is needed (ibid., p. 15).

Is the progress of knowledge, especially rapid advances, indeed a burden on an inclusive democracy, civil society and the capacity of the individual to assert his or her will? If there is a contradiction between knowledge and democratic processes, is it a new development, or is the advance of liberal democracies co-determined by the joint force of knowledge and democratic political conduct that enable one to claim that civil society, if not democracy, is the daughter of knowledge? Is it perhaps a naive faith in knowledge that propels such a conviction?

In short, the tension between what would appear to be the growing utilisation of specialised knowledge in governance in knowledge societies and democratic governance is the theme of our contribution. We cannot claim to have a solution to a conceivable irreconcilability between democracy and specialised knowledge. The two problems posed by moderate critics cannot be solved and deserve particular attention in any study on climate politics. However, we reject the fundamental criticism denying democratic institutions any problem-solving capacity. This means we reject the thesis of an inconvenient democracy.

Throughout modern history, one encounters assertions about a withering away of politics and the substitution for the reign of power of men over men with the authority of scientific knowledge. Without identifying himself with the position in question, the economist Frank H. Knight (1949, p. 271) refers to a naive positivistic conception of the relation between scientific knowledge and societal problems that is repeated many times in the context under discussion: 'Science has demonstrated its capacity to solve problems, and we need only understand that those of the social order are of the same kind.'

With the emergence of urgent global environmental problems, a new vision or a recall of an old vision for the role of scientific knowledge in political governance is becoming evident. The grand vision for the new political role of scientific knowledge is, in turn, linked to a broad disenchantment about the practical efficacy of democracy; the conviction that the public is unable to comprehend the nature of the problems faced by humankind, but also a misconception about the societal role of knowledge, in particular scientific knowledge.

As a result, convictions expressed about the fundamental deficiency of democratic governance – in light of the profound problems humankind faces and must deal with – stand in essential contradiction to another form of alarm and strong doubt expressed about threats to democracy posed by experts, the very experts who warn humankind and policy-makers about the dangers to modern societies by global warming (see Stehr 2016, p. 202).

There is a parallel justification for 'the power of superior (objective) knowledge' and the legitimacy of decisions supported by and derived from such knowledge. Part of such a justification is, for example, a specific understanding of the function of the institution of the state. The famous French sociologist Emile Durkheim refers to this convergence of political legitimacy and knowledge when he remarks:

> If the State does no more than receive individual ideas and volitions to find out which are more widespread and 'in the majority', as it is called, it can bring no contribution truly its own to the life of society.... The role of the State, in fact, is not to express and sum up the unreflective thoughts of the mass of the people but to superimpose on this unreflective thought a more considered thought, which therefore cannot be other than different.
>
> (Durkheim [1950] 1992, p. 92)

Contemporary considerations in the 1920s and 1930s about science, and the adequacy and capacity of democratic governance to cope with the rapid advance of scientific knowledge, the accumulation of urgent societal problems, the spread of totalitarian governments and the rapid rise of the complexity of the world as a result of the growing size of the population, resonate with today's discussions about the global environmental problems and the capacity of democracy to adequately respond. Some scientists in those days, not only Marxists, were prepared to accept, even urge, a stronger regulation of society in the face of massive social and economic problems and hence are prepared to sacrifice some democratic rights.

Today, activist climate scientists, politicians and many other observers agree that the previous climate summits in Copenhagen, Cancun, Durban and Warsaw were failures. The summits did not result in a new global agreement to cope with the emissions of greenhouse gases. In the aftermath of these failures, the summit in Paris in the fall of 2015 was increasingly described as a decisive moment in global climate politics. Stakes were high, as were the fears of a renewed failure:

> It's Paris or bust. Climate diplomats are preparing for a United Nations climate conference in the French capital in December that scientists say is probably the last realistic chance for the world to prevent global warming going beyond 2 degrees Celsius.
>
> (Pearce, 2015)

Accordingly, many observers breathed a sigh of relief when the 'Paris Agreement' was adopted on 12 December 2015. The *Guardian* celebrated the agreement as the 'world's greatest diplomatic success' (Harvey, 2015).

If Hollywood had scripted the Paris negotiations, the successful adoption of the Paris Agreement would indicate not the happy ending, but the moment when scepticism is cast aside and the hero gets the phone call.

An inconvenient democracy? Maybe

Unfortunately, the Paris Agreement might not even indicate a turning point in global climate politics at all. Donald Trump's announcement that the United States would withdraw from the Agreement in June 2017 provoked criticism of the decision and declarations of commitment to fighting human-made climate change from inside and outside the US. Jerry Brown, Governor of California, a strong supporter of climate policies, ranted about Trump's decision even before it was officially announced, adding that 'you can't fight reality with a tweet' (Chaitin, 2017). Likewise, the newly elected French President Emanuel Macron, paraphrasing Trump's campaign slogan, defended the Paris Agreement by issuing a call to 'Make our Planet Great Again' (Abrams, 2017).

Yet it remains to be seen whether the Trump administration's solo-run indicates just a reshuffling of climate alliances, with China and potentially India stepping in and accepting bigger roles in climate politics, or whether the withdrawal of one of the biggest polluters will put the whole international climate agenda in jeopardy. One must not forget that a mere two months after the agreement was adopted; that is, long before Trump took office, the tone already began to change:

Euphoria about the diplomatic success gave way to skepticism if the deal will actually have real political power to initiate ambitious climate policy worldwide that can prevent dangerous levels of climate change.

(Brauers & Richter, 2016, p. 1)

Brauers and Richter continue to analyse the conditionalities of the Paris Agreement (Brauers & Richter, 2016, pp. 2–4). They conclude that the Paris Agreement 'should not be seen as reaching the goal, but rather as a starting signal for an increase in the global ambition to prevent climate change to reach dangerous levels' (ibid., p. 5). One reason for the somewhat cautious optimism is the 'question of legal bindingness' (ibid., p. 3) and the problem of 'objectively' judging the quality of expert advice. In short, the two problems formulated by moderate critics remain.

The two dimensions crystallise when it comes to the future role of markets. It is decisions on energy resources that will fuel economic development. Likewise economic expertise – in favour of or opposing decarbonisation – and climate expertise remain likely battlegrounds for future decision-making on global climate politics.

We can glean some insight into these questions via the work of the renowned American economist and political scientist Charles E. Lindblom (1995), who examined the complex interrelations between knowledge, markets and democracy. These interrelations are just as relevant today, and not just because of the serious effects of the recent financial and economic crisis.

As anthropogenic climate change aptly demonstrates, the supposed virtues of a free *market* can easily be questioned. Apparently unrestrained liberal markets couldn't prevent global greenhouse gas emissions from mounting, and creating problems that might well spiral out of control. Unsurprisingly, policy-makers and climate scientists alike have made proposals to fence market freedom in order to limit its ability to externalise the costs for environmental problems, the most prominent example being carbon pricing (Kossoy, Peszko, Oppermann, Prytz, Klein, Blok et al., 2015). Another example of pronounced scepticism towards the virtues of the 'invisible hand' can, for instance, be found in the *Fourth Carbon Budget*: Reducing Emissions through the 2020s (Committee on Climate Change, December 2010) of the official UK Committee on Climate Change (CCC). Commissioned to advise on how to reach the ambiguous target in reducing emissions, the Committee expresses its doubt about the capacity of an 'unrestrained' market to bring about such goals. It therefore advocates the return to elements of 'political planning'. The report states with respect to the electricity market of the future in the UK that:

[c]urrent market arrangements are highly unlikely to deliver required investments in low-carbon generation.... Given the need to decarbonise the power sector and the long lead-times for low-carbon investments, reform of the current market arrangements to include a system of tendered long-term contracts is an urgent priority.

(Committee on Climate Change, 2010, p. 239)

It must be noted that the Committee takes no anti-market positions, but rather argues in favour of steering the (electricity) markets towards more sustainable modes of production. Critics might notice the striking similarity between the steering and restricting of the markets and the provision of subsidies for industries that are likely to be affected by climate politics. It remains questionable whether democracies would be capable of enforcing policies *against* markets to force climate-friendly transition.

Moreover, much less common is an open and explicit expression of doubt about the virtues of democracy. In particular, it has traditionally been the case that scientists have rarely raised serious misgivings in public about democracy as a political system.

But could this change? What if the Paris Agreement fails to be the much-expected breakthrough? What if the example of the US encourages other countries to deviate from the political guidelines and targets agreed upon in the Paris Agreement? Such developments can become increasingly problematic for democracies. When climate politics are easily thwarted by the (impulsive) political decisions of *elected* representatives and climate policy change that could come a whole election cycle away, trust in the capacity of democratic institutions to deliver timely and lasting solutions may be weakened.

In such a situation, the above-mentioned David Shearman and Joseph Wayne Smith might reaffirm their claim that '[w]e need an authoritarian form of government in order to implement the scientific consensus on greenhouse gas emissions' (cited in Grundmann & Stehr, 2012, p. 182) and reinforce their conclusion that 'humanity will have to trade its liberty to live as it wishes in favor of a system where survival is paramount' (Shearman & Smith, 2007, p. 4). Mark Beeson takes the argument even further and adds that 'forms of "good" authoritarianism, in which environmentally unsustainable forms of behaviour are simply forbidden, may become not only justifiable, but essential for the survival of humanity in anything approaching a civilized form' (Beeson, 2010). Should we conclude that authoritarian states such as China, for example, will become the role models of authoritarian environmentalism (Gilley, 2012)?

Indeed, it remains to be seen whether the Paris Agreement will be able to reconcile frustrated climate scientists like James Lovelock. In his 2009 *The Vanishing Face of Gaia*, he emphasises that we need to abandon democracy in order to meet the challenges of climate change head on. We are in a state of war. In order to pull the world out of its state of lethargy, a Churchillian global warming speech offering 'nothing but blood, toil, tears and sweat' (Lovelock, 2009, pp. 31–32) is urgently needed.

The dominant political approach concentrates almost to the exclusion of other forms and conditions of action, on a single *effect* that governance ought to achieve, namely a reduction of greenhouse gas emissions. By focusing on the goals of political action rather than its conditions, the contentious issue of climate change is reduced from a socio-political issue to a scientific or technical one. In consequence climate politics become depoliticised (that is, the scope of political alternatives is reduced to 'right' and 'irresponsible' decisions respectively), which in turn leads to a politicisation of climate science (since scientific findings compel political action). Sheila Jasanoff (2012, p. 1) asserts that support for such a conversion comes with the societal ascendency of science and technology generally: issues 'that matter to the public have been prematurely taken out of politics'. Another significant outcome of the depoliticisation of climate discourse is to more or less openly insist and declare that the social sciences are irrelevant to climate discourse.

An inconvenient technocracy? Without a doubt

Democratically organised societies are too cumbersome to avoid climate change; they do not act in a timely fashion, nor are they responsive in the necessary comprehensive manner. The 'big decisions' that have to be taken in the case of climate change require a strong state. The endless debate should end. We have to act – that is the most important message. What better solution than to hand over decision-making to experts, right?

In this final section we outline that democracy, as inconvenient and frustrating as it can sometimes be, is always preferable to expert rule and technocratic decision-making. We will not root the superiority of democratic processes in normative assumption, but will link it to specific response capacities built in the democratic system. To develop our argument, we will compare technocratic and democratic decision-making along three dimensions reflecting the alleged shortcomings of democracy.

First, we are informed that, in recent years, the robustness and the consensus in the science community about human-caused climate change has not only increased in strength, but a growing number of studies point to far more dramatic and long-lasting consequences of global warming than previously thought. Moreover, it is highly likely that the sophistication and depth of our knowledge about global and regional climates will substantially increase in the next decades. Under such circumstances, how is it possible, many scientists ask, that such evidence does not motivate political action in societies around the world?

Could this problem be averted in a technocratic system? The answer is no, for the problem of selecting the 'right' expert would remain. Over the past decade, publications by climate sceptics and efforts to undermine climate science have *also* increased (Oreskes & Conway, 2010; Dunlap & Jacques, 2013). The (in)famous climate deniers of the Heartland Institute issue policy papers denying that climate change is a problem and claiming that there is no consensus among climate scientists (see: www.heartland.org/issues/environment). At the same time the Nongovernmental International Panel on Climate Change (NIPCC) seeks to establish itself as a reliable source for scientific expertise by publishing reports which find that 'Nature not human activity rules the climate' (see: http://climatechangereconsidered.org/about-nipcc/).

Outside of the US, these attempts to establish climate-sceptic counter-expertise have not been very successful. However, the selection bias towards 'mainstream' climate science is by no means a 'natural' result. Instead we have to identify the *selection problem* as the main weakness of any technocratic system. It is true that scientific experts have to compete and sometimes fail to gain acceptance in democratic systems. It is also true that the process of expert selection and access to a wider public can be influenced (sometimes heavily) by elite and economic power (Oreskes & Conway, 2010). However, expert selection in technocracies is completely arbitrary. Even if 'scientific standards' are invoked there is no truly objective selection. A short example illustrates this aspect: solving a financial crisis by handing over to economic experts clearly would not do the job. To which expert should we turn? A representative of neo-classical economics? A neo-Keynesian? A neo-Marxist? It goes without saying that the different theoretical conviction of the expert would lead to mutually exclusive problem definitions and result in dramatically divergent policy advice (e.g. austerity measures, deficit spending or nationalisation of companies and banks). The same is true for choosing the 'right' climate scientist. Despite a general consensus among the international community of climate scientists on the existence and the causes of global climate change, dissenting experts willing to challenge this consensus are conveniently available, especially in the United States (see Oreskes & Conway, 2010).

We therefore argue that the procedural dimension of the selection process, the need to (publicly) justify the selection and legitimise the policies derived from expert advice, is the first and most important response capacity of democratic systems.

Second, the still-dominant approach to climate policy shows little evidence of success. One result of the recent global recession was the unintended and temporary reduction of the increase of CO_2 emissions. However, efforts to overcome the crisis do not foreshadow a fundamental shift in economic policies, e.g. towards post-growth strategies. On the contrary, everything is set in motion worldwide for a resumption of economic growth. Jump-starting the economy means that emissions will rise again. As problematic as this development is for global climate politics, it indicates another response capacity of democratic systems: *participation by a wider public in problem selection and agenda setting.*

Critics of democracy underestimate the fact that modern societies never face one challenge at a time. Instead several problems have to be weighted, balanced and addressed simultaneously. Problem selection and agenda setting cannot be done properly in a technocratic system, because it would pose a second-order selection problem. Instead of having to select experts from within one field of expertise one would have to choose which area of expertise should be activated (whether to deal with the economic or the ecological problem first). Things get even more complicated if one assumes that political challenges can be related or even interdependent, for then experts might disagree on how a set of problems relate to each other. In short, only democracies can meaningfully address the problem of *demand* for expertise. In technocracies, the demand is either identified by the experts themselves ('They tell us that we need them') or is somehow 'given'.

Third, in the architecture of the reasoning of the impatient critics of democracy, one notes an inappropriate fusion of nature and society. The uncertainties that the science of the natural processes (climate) claims to have eliminated are simply transferred to the domain of societal processes. Consensus on facts, it is argued, should motivate a consensus on politics. We would like to argue that one of the alleged weaknesses of democratic decision-making turns out to be one of its biggest advantages: *political competition*. While it is true that political competition tends to be incompatible with scientific claims to truth, it is also true that consensus on facts often enough does not lead to consensus on policy advice. There is range of policy recommendations among climate scientists even if climate sceptics remain benched. This means that even transferring power to experts wouldn't prevent debate about appropriate, preferred or 'necessary' political action. However, it would suspend *public* debate on these issues, seriously damaging the legitimacy and acceptance of political measures.

Conclusion

In this contribution, we outlined and explained the occasional uneasiness of democracies in an era of anthropogenic climate change. Indeed, it is tempting to envision a post-political world to deal with a global environmental crisis. However, post-political governance of climate change is attached to specific conditions: 'Central to this post-political thesis is that there is a consensual humanity encountering a universal human threat and that everyone is facing this together' (Urry, 2011, p. 91) We're back in a Hollywood movie. The inevitable sceptic has just been proven wrong and the camera pans out to Mr President, who is about to finally call the hero. Unfortunately, the post-political thesis, as presented by John Urry, is fundamentally flawed. As we have demonstrated above even if there were universal consensus on facts, there need not be a resulting consensus on politics. Moreover, 'humanity' is far from facing a universal threat from climate change. For example, the 'Environmental Vulnerability Index', developed by the South Pacific Applied Geoscience Commission and the United Nations Environment Programme, shows that vulnerability to climate change varies between countries depending on geographic location and institutional capacities to invest in protective measures (www.vulnerabilityindex.net).

For the time being climate change will remain an essentially political issue. In this contribution we have made the case for democratic decision-making to be the best option to deal with the global climate challenge.

This is not to say that democracy does not have to change in order to cope with climate change. As Timothy Mitchell has aptly demonstrated, '[d]emocratic politics developed, thanks to oil, with a peculiar orientation towards the future: the future was a limitless horizon of growth' (Mitchell, 2011, p. 253). It is likely that this connection cannot be maintained; the democratic vision of the future, which includes limitless fossil-fuel driven growth, cannot hold. However, Mitchell doesn't stop here. He continues to argue that the emergence of this particular vision 'was the result of a particular way of organizing expert knowledge and its objects' (ibid.). Handing responsibility over to experts therefore wouldn't solve the problem. What is needed is a re-organisation of environmental and economic expertise. As we have argued above, democratic systems are best suited to manage this process of re-organisation. Due to their 'built-in' capacities to manage the problem of (first and second order) expert selection and their capacity to guarantee political competition democratic states are in the best position to 'think ahead' (Giddens, 2011, p. 94) The discrete charm of authoritarian environmentalism might be explained by the growing desire for 'quick' and 'decisive' action.

However, it is far from clear that authoritarian regimes would do better in the mid- or long run. One must not forget that Western democracies took the lead in the protection of the ozone layer in the 1980s, forging a mutual agreement in Montreal in 1987. Also, they acted as early movers in the campaign to ban leaded fuels and asbestos. Democratic regimes, at times, have demonstrated their capability to deal with serious threats, forge alliances, implement national relations and agree on international binding regulation. To conventionalise today's China to a role model therefore is both misleading and ill-advised. Instead of accusing democracy, Ulrich Beck (2008) emphasises that the flaws and shortcomings of climate politics can be explained by the predominance of attempts to preserve established ways of living by externalising problems in space (especially to the poor countries in the Global South) and time (postponing the solution until future generations have to deal with the problem). So, instead of favouring authoritarian environmentalism and technocratic solutions, climate science must focus on developing credible narratives on future vulnerability (O'Brien & O'Keefe, 2013, p. 115) to support the development of post-normal, yet democratic, risk management mechanisms.

Climate policy must be compatible with democracy; otherwise the threat to civilisation will amount to much more than just changes to our physical environment. In short, the alternative to the abolition of democratic governance as the effective response to the societal threats that likely come with climate change is more democracy and the worldwide empowerment and enhancement of knowledgeability of individuals, groups and movements that work on environmental issues. 'There is but one political system that is able to rationally and legitimately cope with the divergent political interests affected by climate change and that is democracy' (Stehr, 2015, p. 450).

Notes

1 It was Niklas Luhmann ([1991] 2005, pp. 21–22) who introduced the distinction between dangers and risks. Dangers refer to what individuals are exposed to due to external causes such as environmental conditions; risks refer to losses as the result of decisions individuals made; that is, what people chose to dare.
2 As quoted in Andrew Revkin, 'A Risk Analyst Explains Why Climate Change Risk Misperception Doesn't Necessarily Matter', *New York Times* 16 April 2014.
3 Although 'climate sceptics' are still active in public discourses and the political arena in many countries among them, most prominently, the USA see Oreskes and Conway, 2010.

4 A recent example is the 'eurozone crisis' in which the alleged need for 'immediate and appropriate' responses served as a justification for the deepening of the democratic deficit (Schmidt, 2015) and moves towards more technocratic decision-making (Ruser, 2015b).
5 The climate scientist Hans Joachim Schellnhuber, in an interview with *DER SPIEGEL* (Issue 12, 21 March 2011, p. 29) in response to the question why the messages of science do not reach society.
6 Keller (2010; also 2011) arrives at her conclusion about the inseparability of science and politics and the political authority of climate science by suggesting that 'where the results of scientific research have a direct impact on the society in which they live, it becomes effectively impossible for scientists to separate their scientific analysis from the likely consequences of that analysis'.

References

Abrams, Abigail (2017, 1 June). French President Emmanuel Macron Trolls Donald Trump: 'Make Our Planet Great Again' *Time* http://time.com/4802549/emmanuel-macron-trolls-donald-trump-paris-climate-agreement/ accessed 17 June 2017

Beck, Ulrich (2008). Climate Change and Globalisation are Reinforcing Global Inequalities: High Time for a New Social Democratic Era, *Globalization, 5*, 78–80

Beeson, Mark (2010). The Coming of Environmental Authoritarianism. *Environmental Politics, 19*, 276–294.

Berlin, Isaiah ([1958] 1969). Two Concepts of Liberty. In Isaiah, Berlin *Four Essays on Liberty*. Oxford: Oxford University Press.

Brauers, Hanna & Richter, Philipp, M. (2016). The Paris Agreement: Is it Sufficient to Limit Climate Change?. *DIW Roundup, 91*, 1–6.

Chaitin, Daniel (2017, 31 May). Gov. Jerry Brown: Trump fighting a losing battle with 'reality' on Twitter *Washington Examiner*, www.washingtonexaminer.com/gov-jerry-brown-trump-fighting-a-losing-battle-with-reality-on-twitter/article/2624632, accessed: 11 June 2017.

Committee on Climate Change (2010). *The Fourth Carbon Budget: Reducing Emissions through the 2020s.* London: Committee on Climate Change.

Comte, Auguste (1855). *Social Physics: From the Positive Philosophy of Auguste Comte.* New York: Calvin Blanchard.

Dunlap, Riley, Jacques, Peter (2013). Climate Change Denial Books and Conservative Think Tanks. Exploring the Connection. *American Behavioral Scientist, 5*, 699–731.

Durkheim, Emile ([1950] 1992). *Professional Ethics and Civic Morals.* London: Routledge.

Elias, Norbert ([1987] 2006). Über den Rückzug der Soziologen auf die Gegenwart. In Norbert Elias *Gesammelte Schriften. Band 16: Aufsätze und andere Schriften III.* Frankfurt am Main: Suhrkamp.

Elias, Norbert (1987). The retreat of sociologists into the present, *Theory, Culture & Society, 4*, 223–247.

Giddens, Anthony (2011). *The Politics of Climate Change* (2nd ed.). Cambridge: Polity.

Gilley, Bruce (2012). Authoritarian Environmentalism and China's Response to Climate Change. *Environmental Politics, 21*(2), 287–307.

Grundmann, Reiner & Stehr, Nico (2012). *The Power of Scientific Knowledge. From Research to Public Policy.* Cambridge: Cambridge University Press.

Hansen, James (2010). *Storms of my Grandchildren. The Truth about the Climate Catastrophe and Our Last Chance to Save Humanity.* Bloomsbury Publishing.

Harvey, Fiona (2015, December 14). Paris climate change agreement: the world's greatest diplomatic success. *Guardian.*

Hayek, Friedrich ([1960] 2006). *The Constitution of Liberty.* London and New York: Routledge.

IPCC (2014). Climate Change 2014: Mitigation of Climate Change. *Contribution of Working Group III to the Fifth Assessment Report of the Intergovernmental Panel on Climate Change* [Edenhofer, O., R. Pichs-Madruga, Y. Sokona, E. Farahani, S. Kadner, K. Seyboth, A. Adler, I. Baum, S. Brunner, P. Eickemeier, B. Kriemann, J. Savolainen, S. Schlömer, C. von Stechow, T. Zwickel and J. C. Minx (Eds.)], Cambridge and New York: Cambridge University Press.

Jasanoff, Sheila (2012). *Science and Public Reason.* London: Routledge.

Keller, Evelyn Fox (2010). "Climate science, truth and democracy," unpublished manuscript.

Keller, Evelyn Fox (2011). What are climate scientists to do? *Spontaneous Generations. A Journal for the History and Philosophy of Science, 5,* 19–26.

Knight, Frank H. (1949). Virtue and Knowledge: The View of Professor Polanyi. *Ethics, 59,* 271–284.

Kossoy, Alexandre; Peszko, Grzegorz; Oppermann, Klaus; Prytz, Nicolai; Klein, Noemie; Blok, Kornelis; Lam, Long; Wong, Lindee; Borkent, Bram (2015). *State and Trends of Carbon Pricing 2015.* Washington, DC: World Bank.

Lahsen, Myanna (2005). Technocracy, Democracy, and the U.S. Climate Politics: the Need for Demarcations. *Science, Technology & Human Values, 30,* 137–169.

Lindblom, Charles E. (1995). Market and Democracy – obliquely. *PS: Political Science and Politics, 28,* 684–688.

Lovelock, James E (2009). *The Vanishing Face of Gaia: A final Warning.* New York: Basic Books.

Luhmann, Niklas ([1991] 2005). *Risk.* A Sociological Theory. With a New Introduction by Nico Stehr & Gotthard Bechmann. New Brunswick, NJ; Aldine Transaction.

Merton, Robert, K. (1948). The Self-Fulfilling Prophecy. *The Antioch Review, 8,* 193–210.

Mill, John Stuart (1861). *Consideration on Representative Government.* London: Parker, Son and Bourn.

Mitchell, Timothy (2011). *Carbon Democracy. Political Power in the Age of Oil.* London, New York: Verso.

O'Brien, Geoff & O'Keefe, Phil (2013). *Managing Adaption to Climate Risk: Beyond Fragmented Responses.* London: Routledge.

Oreskes, Naomi & Conway, Erik M. (2010). *Merchants of Doubt.* London, New Delhi, New York, Sydney: Bloomsbury.

Pearce, Fred (2015, September 14). Will the Paris Climate Talks Be Too Little and Too Late? *Environment 360.*

Powell, Norman J. (1951). *Anatomy of Public Opinion.* New York: Prentice-Hall.

Ruser, Alexander (2015a). Sociological Quasi-Labs: The Case for Deductive Scenario Development. *Current Sociology, 63,* 170–181.

Ruser, Alexander (2015b). By the Markets, of the Markets, for the Markets? Technocratic Decision-Making and the Hollowing Out of Democracy. *Global Policy, 6,* 83–92.

Ruser, Alexander (2016). Towards the Unity of Science again? Reductionist thinking and its consequence for a social philosophy of science. *Epistemology and Philosophy of Science, 49*(3), 55–69.

Schmidt, Vivien A. (2015). The Eurozone's Crisis of Democratic Legitimacy: Can the EU Rebuild Public Trust and Support for European Economic Integration? European Commission, Discussion Paper 015, September 2015.

Shearman, David & Smith, Joseph Wayne (2007). *The Climate Change Challenge and the Failure of Democracy.* Westport, London: Praeger.

Snow, C.P. (1959). *The Two Cultures and the Scientific Revolution.* The Rede Lecture. New York: Cambridge University Press.

Stehr, Nico (2015). Democracy is not an Inconvenience. *Nature, 525*, 449–450.

Stehr, Nico (2016). *Information, Power, and Democracy. Liberty is a Daughter of Knowledge.* Cambridge: Cambridge University Press.

Urry, John (2011). *Climate Change and Society.* Cambridge: Polity.

Wildavsky, Aaron (1979). *Speaking Truth to Power. The Art and Craft of Policy Analysis.* New Brunswick and London: Transaction Publishers.

2 The institutional capacity of democracy

Theresa Scavenius

Introduction

Democratic institutions provide a framework within which individuals and groups can govern climate change. There is an increasing concern, however, that the current generation of democratic institutions is not capable of managing climate change satisfactorily. There are several reasons why liberal democracies might be inherently reluctant to mitigate environmental harm and climate change. One answer addresses the question of policy choice. Democracies might have higher or more urgent priorities or they are vulnerable to short electoral cycles, attenuating attention to long-term issues. Some commentators have suggested that a reason for this is that human beings might not have 'the nature equipped with a moral psychology that empowers them to cope with the moral problems that these new conditions of life create' (Persson & Savulescu, 2012, p. 1); moral deficiencies that the political system of liberal democracy might not be able to overcome. Another common suggestion is that democratic institutions are prone to sclerotic lobbying by corporate stakeholders and politicians that benefits them rather than their political constituents (Shearman & Smith, 2007, p. 165; Persson & Savulescu, 2012, p. 1).

In this chapter, I discuss the critique of democratic climate governance by distinguishing between a 'proactionary' and 'authoritarian' critique. Where the latter is clearly incompatible with democracy, I show that the former also presumes concepts of politics and agency that are alien to theories of democracy. Moreover, I show that despite their differences, the two critiques converge in their neglect of the importance of maintaining the democratic capacity of political institutions, such as parliaments and other formal institutions of government.

I argue that the current failures of climate governance are not merely a question of moral capacities, but rather one of deteriorating institutional capacity that renders effective democratic climate governance impossible. The failures of democratic climate governance reflect the deeper challenge of the institutional capacity of our current political institutions. Democracies with low political capacity are less fit to rectify environmental harm and climate change. If democratic societies are currently incapable of implementing comprehensive climate change, I argue that the only plausible solution is not a dismissal of democracy, but political reforms of the current democratic institutions that would reinvigorate their capacity to conduct climate governance and other important policy agendas.

The chapter proceeds as follows: first, I consider the critique of democracy based on ideas of people's lack of moral inclinations and the authoritarian quest. I then argue that an important intellectual resource for understanding the critique of democratic climate governance is to be found in the current underestimation of the intermediating factors and the institutional capacity of contemporary democratic societies to handle climate challenges collectively and politically. 'Institutional capacity' is used here to refer to a political institution's capacity to facilitate, monitor and implement policies in a legitimate, effective and transparent manner (Shue, 1988; Ostrom, 2005; Le Grand, 2006; Olsen, 2007; Searle, 2010; Hurley, 2011; List & Pettit, 2011). I conclude that, instead of encouraging a meritocratic and authoritarian trajectory, democratic reforms are a more reasonable and promising solution to the current climate governance deficit.

Proactionary policies

In his seminal article titled 'The Tragedy of the Commons', Garrett Hardin argues that the collective problems of environmental harm and climate change cannot be solved by technical means – they require novel *political* solutions (Hardin, 1968, pp. 1243–1248). Today, however, the critique has turned around. Scholars, scientists and politicians increasingly question the ability of democracy to mitigate climate change by political means (Shearman & Smith, 2007; Beeson, 2010; Persson & Savulescu, 2012; Liao, Matthew, & Roache, 2012). Instead, technical and authoritarian solutions are suggested. From a time when climate activists talked about an 'inconvenient truth', scientific experts are now increasingly uneasy with the 'inconvenient democracy' (Stehr, 2013, pp. 50–60; see also Chapter 1 in this volume).

One of the inconvenient truths is driven by scepticism about people's *moral* and *social* capacities (Persson & Savulescu, 2012, ch. 2). Studies of psychological and genetic dispositions are invoked to vindicate this type of sceptical understanding of people's capacities. We are told that people are ill-equipped to deal with the challenges of climate change because of their 'Stone Age mindset', which restricts the moral motivation of, for example, paying the costs of mitigating climate change for the benefit of future generations (ibid., pp. 1–2).

Proponents of the critique of moral capacity put no faith in the capacity of democratic politics to compensate for the moral deficiencies of humanity and contemporary democracies (Sherman & Smith, 2007, p. 165; Persson & Savulescu, 2012, pp. 1, 122). They suggest various radical policy solutions, including the so-called 'proactionary' solutions. Proactionary approaches are the opposite of *precautionary* approaches. Precautionary policies aim to prevent the worst outcomes, whereas proactionary policy-makers seek to promote the best available opportunities. Contrary to the worries of the precautionary camp, the proactionary approach is to encourage people to transcend current norms rather than adhere to them. Proactionary policy suggestions experiment with new technologies and new pharmacological prospects (Fuller, 2012, pp. 1–2).

While proactionary policies are not inherently incompatible with democratic climate governance, the underlying understanding of people's lack of moral competences provides the justification of proactionary policies, which render democratic climate policy-making redundant. For example, Persson and Savulescu suggest the controversial and radical 'medical moral enhancement', which is supposed to augment empathy and sympathetic concerns about the well-being of others and future beings (Persson & Savulescu, 2012, p. 109). Two drugs are suggested as relevant to moral bioenhancement: the hormone and neuorotransmitter oxytocin and glucocorticoids. Studies of the influence of these drugs on moral sentiments are not conclusive; some studies even demonstrate the possibility of the drug's promoting anti-group moral sentiments. Nonetheless, the authors argue that we should not preclude biological moral enhancement as a possible way to reverse neglected climate action and to regain control with climate governance.[1]

Another type of bioengineering aims to change people's incentives for eating red meat. One big problem of climate change is the methane gas released from livestock farming, which is estimated to account for at least 37% of the world's greenhouse emissions. Thus, a reduction in livestock farming would have a positive environmental impact.[2] The idea is to induce a pharmacological meat intolerance, i.e. a vomiting substance, to manipulate people's excessive desire for red meat. Adding a vomiting substance to red meat will make people voluntarily avoid the environmentally unfriendly meat, causing a net benefit for human beings and nature (Liao et al., 2012).

Yet others defend a modification not of human moral or bio-capacities, but of the earth's ecosystem. Several geoengineering strategies have been suggested. One much-discussed idea suggests a modification of solar radiation by injections of sulphur dioxide gas into the stratosphere (see Chapter 7 in this volume for further discussion on geoengineering).[3] Liao et al. (2012) argue that human engineering may be less risky than geoengineering. Moreover, they emphasise, as do Persson and Savulescu, that equivalent pharmacological and biomedical forms of human enhancement and modification are already available. On the basis of this, they conclude that opponents of human engineering do not have a strong case.

Despite their differences, proponents of bio- and geoengineering of human and environmental conditions share a radical willingness to engage in proactionary policies, or at least to engage in research about radical solutions. Some proponents, such as Persson and Savulescu (2012) and Shearman and Smith (2007), argue that the urgency of climate change compels us to experiment with both human and environmental conditions, possibly disregarding the moral integrity of mankind and the earth, and therefore also the moral integrity of democratic political institutions.

Several points need to be considered in order to evaluate such proposals on moral grounds. First, one should consider how the policies are justified. One way of justifying the proactionary policies is voluntarism. The policy of inducing meat intolerance is claimed to be a voluntary practice. Liao et al. argue that 'human engineering would be a voluntary activity [...] rather than a coerced, mandatory activity' (Liao et al., 2012, p. 211). Their argument for voluntarism, however, rests on thin ground: it belittles or neglects several important concerns. They argue that human engineering is necessary because 'people often lack the motivation or willpower to give up eating red meat even if they wish they could' (ibid., p. 212). However, if individuals lack motivation and willpower with regard to climate action, consumers cannot be expected to have the willpower to choose the modified meat instead of normal meat or no meat at all. Second, other proponents of proactionary policies do not even presume voluntarism. Persson and Savulescu argue that individuals may be subjected to the proactionary policies without their consent:

> [S]ome children should be subjected to moral bioenhancement [...]. This is because the capacity to influence development under way is likely to be greater than the capacity to alter established motivational dispositions and behaviour. There is no reason to assume that moral bioenhancement to which children are exposed *without their consent* would restrict their freedom and responsibility more than the traditional moral education to which they are also exposed without their consent.
>
> (Persson & Savulescu, 2012, p. 114, emphasis added)

Three further points should be added. First, this argument slides towards an authoritarian argument, which will be discussed below. Second, if human nature is changed by pharmacological means, it is uncertain to what extent it makes sense to talk about voluntarism of human agents in the future.[4] If we change the nature of humanity fundamentally, we need to reconceptualise what human agency means. Moreover, policies that disregard the moral integrity of human nature, human rights and the moral value of democratic participation are unlikely to respect voluntarism.

Not all proactionary policy suggestions change people's nature. But even if the nature of human agency is not changed, other problems remain. For example, how are these proactionary policies to be implemented? *Who* should decide *when, how* and *whom* to subject to these polices? Little is said about how these policies should be governed. Scholars of geoengineering have developed a set of so-called 'Oxford Principles' to guide geoengineering research and politics. A key point of these principles is that geoengineering requires public consent and participation. Another principle states that geoengineering research should be based on open access to data results and independent assessment studies (Rayner, 2009; see also Chapter 7 in this volume). These principles are valuable, but may be insufficient to overcome the limited options of participation, transparency and openness in climate disaster situations, which many proponents of geoengineering envisage. While some advocates of geoengineering experiments base their case on an urgent need to avoid climatic catastrophe that could trump democratic considerations, Rayner and Healey (Chapter 7, this volume) argue that an institutional architecture for geoengineering research needs to be developed in order to secure societally legitimate management of geoengineering research and experiments.

Moreover, human- and geoengineering are likely to be large-scale projects in need of enormous investments. Robock argues that commercial and military interests are likely to control such projects (Robock, 2008, p. 17). Hence, although the proponents of proactionary policies maintain the ideal of voluntarism, benevolence and democracy, there are several reasons why proactionary policy programmes may be ill-suited to maintaining voluntarism and strong democratic participation. Indeed, there is a danger that it will take authoritarian rule to enforce the suggested biological modifications of people's moral capacities and motivations and the deliberate geological modifications of the earth's climate and nature, such as the solar radiation technologies. Therefore, it also does not come as a surprise that the argument for proactionary policies can slide towards the argument for a benevolent dictator, which will be discussed in the next section.

The authoritarian solution

Several proactionary scholars explicitly embrace different types of 'good authoritarianism' as a way to restore political control with climate politics (Sherman & Smith, 2007; Lovelock, 2010).[5] Nonetheless, it is relevant to distinguish the proactionary policy suggestions that aim to change people's motivations from the authoritarian policy suggestions. The reason for this distinction is that the two aspects rely on two rather different and, as we shall see, not fully compatible sets of assumptions, and therefore also on two different critiques of contemporary democratic climate governance.

The point of departure of the proactionary critique, discussed above, takes a sceptical view of human nature: human nature makes individuals unlikely to do something good for the climate and people's selfish interests make political institutions inherently unable to implement the relevant policy options necessary to curb climate change.[6]

Let us now look into two of the arguments for authoritarian proactionary policies. Proactionary policies and authoritarian policies share a common premise that there exist two types of human beings. The idea is that some unfortunate people have a 'less morally motivated nature' and other more fortunate people have a morally motivated nature (Persson & Savulescu, 2012, p. 113). The problem in democracy is that those in political power are not necessarily the wise and the benevolent ones (see e.g. Shearman & Smith, 2007). Hence, in order to restore democratic climate politics, it is not enough to replace democratic decision-making with voluntary proactionary policy frameworks. In contrast, proponents of authoritarian proactionary policies consider a *benevolent dictator* or *a meritocratic elite* as the only options. Democracies are thought to lack control of climate politics because they allow too many unwise and malevolent people to influence politics, which slows and ultimately derails the decision-making process.

Proactionary policy proposals can be implemented within democratic decision-making institutions, but they remain alien to democratic values because of their sceptical view on human nature. The authoritarian critique suggests more directly *anti-democratic* climate governance. The benevolent dictator or elite is the answer for those who have lost patience with 'the seemingly interminable machinations of polycentric politics' (Dryzek & Stevenson, 2011, p. 1865). One scientist and climate activist proclaims that

> even the best democracies agree that when a major war approaches, democracy must be put on hold for the time being. I have a feeling that climate may be an issue as severe as a war. It may be necessary to put democracy on hold for a while.
>
> (Lovelock, 2010)

The claim is that the democratic form of governance is incompatible with the governance of climate disasters. A benevolent dictator or elite is thought to be better capable of managing and controlling the current state of affairs in an environmentally and climate-friendly way.

The eco-authoritarians of the 1970s had a similar argument. Some democratic rights and rules 'would have to be sacrificed to achieve sustainable future outcomes since authoritarian regimes are not required to pay as much attention to citizens' rights in order to establish effective policy in key areas' (Held & Hervey, 2009, p. 5).[7] As Ophuls argues, 'scarcity in general erodes the material basis for the relatively benign individualistic and democratic politics characteristic of the modern industrial era' (Ophuls, 1977, p. 163). These anti-democratic arguments are not justified by voluntarism, as some of the proactionary arguments mentioned above are, but by the *benevolence* of the authoritarian regime. In other words, it is assumed that the authoritarian regime will rule in favour of the benefits for all and for nature. One suggestion is that the authoritarian regime should be governed by a well-educated intellectual elite class with a special university degree (Shearman & Smith, 2007, Ch. 9).

The factual aspects of the critique

So far, examples of the critique of democratic climate governance have been presented, and the *proactionary* critique and *authoritarian* critiques of democratic climate governance have been distinguished. The two critiques of democracy are not proposed as normative critiques of democratic institutions and values. In contrast, they are justified as second-best options given the fact that democratic institutions have proven to be incapable of delivering satisfactory climate policies. The necessity of critique is defended in the factual circumstances of politics in order to address the current failures of democratic climate governance. However, the factual components of these critical points have not been adequately defined.

In this section, I argue that we might agree on the factual components of the critique of democratic climate governance without accepting the necessity of initiating proactionary or authoritarian policy experiments.

The critique of democratic climate governance takes its point of departure in the factual analysis that current 'Western societies are not democracies as such but plutocracies, societies ruled by the wealthy', and that 'traditional liberalism has been too permissive as regards letting citizens of affluent societies adopt ways of living that waste the resources of the planet' (Shearman & Smith, 2007, p. 91; Persson & Savulescu, 2012, p. 122). One reason for this is that there exists

> a multitude of forces that are acting to corrupt them and prevent them from truly representing the voice of the people. Various powerful elite groups rule the modern liberal democracy, being based in finance, media, business, and the military; and they have their own agenda that does not have advancing the interests of the common good among them.
>
> (Shearman & Smith, 2007, p. 89)

Further facts support this critique. Indeed, several empirical studies of the current status of democratic quality support the elitist component of the authoritarian critique of democratic climate governance (Streeck, 2011, p. 24). Recent studies show that the quality and transparency of contemporary democracies is deteriorating, the number of 'full' democracies is decreasing, and the strength of the full democracies is deteriorating (Economist Intelligence Unit, 2015). France, Italy, Greece and Slovenia have fallen from the category of full democracies to the category of flawed democracies, a category that includes South Africa, Botswana, Columbia and Thailand (Economist Intelligence Unit, 2011, pp. 5–10).[8]

Yet other empirical studies on the correlation between democracies and environmental outcome are non-conclusive. Democratic societies do not unequivocally have a better climate action record than non-democratic societies. Some results show that there is a strong correlation between democracy and high environmental quality.[9] Other results are less conclusive. Midlarsky, for example, finds that democracies have good performance in land area protection but not in deforestation, CO_2 emissions and soil erosion (Midlarsky, 1998).

The critics of democratic governance face several problems, however. First, there is a discrepancy between the often harsh and strongly worded critiques and the very few lofty ideas for solving the problem of motivation or incentives (Persson & Savulescu, 2012, p. 1; Shearman & Smith, 2007, p. 165). No comprehensive sustainable policy design is suggested. Persson and Savulescu (2012) and Liao et al. (2012) suggest pharmacological treatments that do not yet exist and which, they themselves admit, are unlikely to exist in the near future. Shearman and Smith (2007) provide an underdeveloped idea of fostering new ethical education of the meritocratic elite (Shearman & Smith, 2007, p. 152).

A second problem is that the different critiques converge in their neglect of the importance of maintaining the political institutions. Their critical positions fail to provide a comprehensive understanding of politics, governance and institutional capacities that is competitive with the political theories of democratic institutional design and decision-making processes. They fundamentally underestimate the importance of including ideas about how the institutions upon which their policy recommendations rest are maintained. Although they despair of democratic processes, and emphasise democracy's institutional challenges, they exhibit a naive faith in authoritarian technocracy, which, in reality, invariably presents its own institutional pathologies of tunnel vision and corruption. The proactionary critique presumes the existence of a political system that is willing and has the capacity to conduct trans-human and trans-geological experiments in a voluntary and benevolent way. And the authoritarian regime suggestion presumes the existence of a well-educated meritocratic elite class that has the will, capacity and power to administer and rule society by environmentally friendly means. Nothing is said about how to maintain the social, political and institutional conditions of proposed climate policies.

In contrast, democratic theory is not merely a theory of decision-making processes, nor is it merely a technological fix (Streeck, 2011). It is an epistemic theory of what collective groups of people are capable of doing if they coordinate their acts and behaviours collectively and institutionally, and a political theory with dynamic and flexible instruments to sustain and enhance its institutional design and capacity.

Hence, if it is true that democratic institutions are currently corrupted by corporate and elitist interests, as the critics argue, the critique of democratic climate governance must be accepted, not for normative but for factual reasons. However, it remains to be discussed what conclusions we might draw from this. An alternative to proaction and authoritarianism would be to reinvigorate democracy and develop the institutional capacity of contemporary political societies.

Democratic reforms

The recent history of democratic governance does not univocally provide reasons for the sceptical view of human nature underlying the critique of democratic governance discussed above (Le Grand, 2006, p. 4). There are many examples of successful national and international cooperation on collective action problems of security, environmental matters and public infrastructure and services, such as, at the international level, the Treaty on the Non-Proliferation of Nuclear Weapons (1970) and the Montreal Protocol (1989) and, at the national level, social security systems and educational institutions. These successes indicate that there is no simple correlation between one democratic institutional design and the success and effectiveness of climate politics. One reason for this is that people's behaviour varies according to institutional designs and behavioural spaces. Recent studies have confirmed this (Frey & Jegen, 2000; Ostrom, 2005; Le Grand, 2006, Ch. 3).

Le Grand argues that people sometimes act according to economic motivation and sometimes according to moral motivation. He concludes that financial rewards for something people are morally motivated to do may 'erode the magnitude of the sacrifice that he or she is making and thereby partly erode the motivation to act' (Le Grand, 2006, p. 67). Le Grand does not argue that motivational structure may not be part of policy packages, but that politicians should be aware of whether a given policy is designed to align economic or moral motivation. Only then would it be possible to identify whether one economic or moral incentive design should be implemented or not.

Specifically, it is important to allow for a political system that is capable of channelling people's moral motivations with regard to climate change through democratic decision-making processes. Of foremost importance is the fact that the current problems that democratic climate governance faces may be caused by climate politics that focus primarily on the agent's economic incentives, which potentially erode the relevant agents' ability to act at all. So conceived, the only way out of a climate disaster is not through pharmacological means. Rather, the imperative is to enhance and improve the democratic institutional capacity through which people's moral inclinations can flow freely (Scavenius, 2014).

The democratic theory discussed here takes as its point of departure democratic theory on how free education, science, deliberation and representation allow for knowledge-producing social engagements (Scharpf, 1973; Rostbøll, 2008; Ansell, 2011). In this section, I elaborate the conceptualisation of why deliberation within democratic institutions enhances the social level of knowledge. The argument I pursue is an epistemic argument that knowledge becomes more robust statistically in institutional designs that are capable of producing and advancing knowledge (Popper, 2002; List & Pettit, 2011).

The reason for this is that open and free societies are better at avoiding mistakes and misinformation by increasing the number of free and competent people engaged in scientific and non-scientific evaluations and assessments (List & Pettit, 2011, p. 90). This is not an uncontested argument; many scholars have criticised the knowledge-sharing problems of democracies (List & Goodin, 2001). Notwithstanding this, I find the argument sustains some plausibility.

Based on the epistemic argument, democratic majority voting and deliberation outperforms both dictatorial and unanimity rules in terms of maximising the group's reliability on a given proposition. The premise for this is that people are (1) independent, (2) fallible and (3) biased towards the truth in their judgement (List & Pettit, 2011, p. 90). This means that the agents participating in the vote and in deliberation should be allowed to vote and express their opinions independently of other agents, and moreover that the agents are making fallible assertions which nonetheless are assumed to be biased toward the truth.

To be more precise: when a group of independent and competent people seeks to track the truth, democratisation of the process increases the epistemic gains (List & Pettit, 2011, p. 90). Consider one example. Scientist Francis Galton investigated a contest in which 800 participants were asked to estimate the weight of an ox on display. While few participants were individually accurate, the average estimate across the group of participants turned out to be 1,197 pounds, almost identical to the true weight of 1,198 pounds (List & Pettit, 2011, p. 86).

There is a statistical explanation for why there is an epistemic gain from taking the average of a large number of independent and fallible judgements.

The point is that the likelihood that *one* judgement is accurate is very low. In contrast, the likelihood is very high that the *average* of all judgements is accurate. Moreover, accuracy increases if the contest is repeated, taking the average of a new collection of judgements. In other words, the average of the average has a very high likelihood of being very accurate. This type of information pooling is harnessed in democracies in which each citizen is allowed to vote on ballots and in referendums, participate in public debates and influence politics in other ways.

Further, if the conditions for information pooling in democratic institutions are the independence and competence of the participants (List & Pettit, 2011, p. 87), we can conclude that threats to the democratic politics of climate change are illiteracy, underinvestment in basic education, and the lack of free access to scientific results and deliberation in the media. People can only be competent to engage in democratic climate governance if their access to free media, education and science is secured.

In other words, there is a correlation between the democratic institutional design of free media, education and science on the one hand, and a society's ability to address climate change politically on the other hand. How well a democratic society can address climate change correlates with how well the organisational structure can make use of the aggregated epistemic gains that people collectively mobilise. According to this theory, centralisation of the political process will decrease the epistemic gains.

This claim resembles Friedrich Hayek's famous critique of centralised political planning (Hayek, 1973). In the same way as the market benefits from independent and competent participants with regard to valuing market goods, political democracies benefit from independent and competent citizens with regard to tracking the best policy options. Notice, however, that the two types of aggregation and collection of information differ fundamentally. Whereas the collection of market information is based on a fixed understanding that competition and economic incentives drive people's choices, democratic information pooling is based on independence, competence and bias towards the truth.

More often than not, however, democratic decision-making and deliberation are not based on these three criteria. Sunstein has pointed towards the 'informational' and 'reputational' challenges of democratic deliberation, where participants either 'withhold' important information or 'imitate' participants with higher social reputation (Sunstein, 2007, pp. 13–17). Both types of challenges or negative cascade effects raise serious doubts about the 'miracle' of democratic aggregation (Landemore, 2012, p. 12). Likewise, scholars defending the 'wisdom of the crowd' arguments contend that markets may outperform political processes of aggregation (Surowiecki, 2004). The reason why the market avoids the informational and reputational cascades is because it is a non-deliberative form of aggregation. The market is also thought to function well because people are awarded if they or their groups perform well (Sunstein, 2007).

The statistical explanation of epistemic gains in democratic deliberation and aggregation works only if the participants are not subjected to informational and reputational cascades that make them less likely to perform independently and competently. What we can conclude from this is that democratic deliberation should not be discarded in favour of authoritarian societal models. It is essential for well-functioning social and political institutions that people are independent and have the socio-economic and cultural resources to act freely.

Thus, before democratic institutions are discarded as unfit for climate governance, democratic reforms of the current democratic environment should be considered as possible strategies to restore the political capacity of these institutions to implement climate policies. Even if one accepts the factual aspects of the proactionary and authoritarian critiques – that many contemporary democracies have failed to govern the climate satisfactorily and no longer are *real* and *full* democracies – an alternative inference may be the necessity for democratic reforms of political elitism. The democratic critique stresses that democratic politics has become too elitist and that elitism undermines the institutional, deliberative and participatory advantages of democratic decision-making. The problem of political elitism should not be confused with the issue of expertise. The elitist argument discussed here addresses the point that a well-functioning decision-making process requires a well-functioning political culture of participation, deliberation and open channels for expressing independent, competent and sound judgements.

There is no easy fix for the current challenges democracies are facing, however. Simply providing more and better scientific information making does not guarantee open and democratic debate, let alone sound policy-making. For example, the US has been the leading producer of climate science, yet it is also the origin and centre of climate denial, which has proven resistant to extensive campaigns of science communication insisting that '97% of scientists agree' that climate change is a serious human-caused problem. Social science research has shown that climate denialism is not the result of lack of access to information or cognitive failure, but a manifestation of dysfunctionally polarised political identity that is exacerbated by the activities of partisan think tanks (Lahsen, 2005; Stehr, 2013). On the other hand, recent work on US agricultural producers demonstrates that farmers adapt their agricultural production to climate change, although they present themselves as climate deniers (Rejesus, Mutuc-Hensley, Mitchell, Coble, & Knight, 2013). This suggests that there could be room for engagement across the political divide if the framing of the debate can be shifted to accommodate the institutional identities of both sides.

The democratic argument states that rather than replacing democracy with a more authoritarian governmental authority, fundamental democratic reforms should be considered as a way to fortify and invigorate the democratic quality of flawed democracies. People's independence and competence to make truth-biased judgements should be protected by securing access to free media, research and education. This would arguably also enhance the fitness of contemporary democracies to provide climate governance.

Political institutions do not become more fit to govern the climate by centralising the political decision-making process in elitist structures, or by trespassing on the moral integrity of humanity, as suggested by proponents of the critics of democratic climate governance. In contrast, political institutions can only manage the problems related to environmental harm and climate change if they are well-functioning democratic institutions that can reap the epistemic benefits of democratic institutionalised decision processes, deliberation and participation.

Conclusion

In this chapter, I have discussed the proactionary and the authoritarian critiques of democratic climate governance. Although there is some empirical evidence supporting a critique of democratic climate governance, such critiques suggest policy strategies that are ill-founded. In contrast, I have argued that the failures of climate governance are caused by the deteriorating institutional capacity of democracies, which renders effective democratic climate governance impossible. Democracies with low political capacity are less fit to rectify environmental harm and climate change.

If democratic societies are currently incapable of introducing comprehensive climate politics, dismissal of democracy is not the only plausible alternative. A more obvious strategy is political reforms that re-democratise the political institutions that advance the institutional capacity of societies to conduct climate governance. A key challenge is to find new ways to reap the benefits of free research, media, freedom of expression and dynamic political processes. Currently, this is particularly difficult when democratic governance is becoming increasingly elitist and plutocratic. The failures of democratic climate governance reflect a deeper challenge of a declining institutional capacity of our current political institutions – a challenge that is very difficult to circumvent.

Notes

1 See Persson and Savulescu 2012, p. 133; see also Jefferson, Douglas, Kahane, and Savulescu, 2014. The pharmacological treatment is thought to be especially necessary for 'aggressive males', based on the assumption that women have a greater biological capacity for empathy than men (Persson & Savulescu, 2012, p. 111). One reason for this is that the hormone oxytocin is produced naturally during birth and breastfeeding, thereby mediating maternal care and other pro-social attitudes (ibid., p. 118). If this is true, one plausible inference is not to enhance the anti-social attitudes of men, but to hand over political power to breast-feeding women. This argument resembles the gist of *ecofeminism*.
2 As Liao et al. suggest,

> while reducing the consumption of red meat can be achieved through social, cultural means, people often lack the motivation or willpower to give up eating red meat even if they wish they could. Human engineering could help here. Eating something that makes us feel nauseous can trigger long-lasting food aversion. While eating red meat with added emetic (a substance that induces vomiting) could be used as an aversion conditioning, anyone not strongly committed to giving up red meat is unlikely to be attracted to this option. A more realistic option might be to induce mild intolerance (akin, e.g. to milk intolerance) to these kinds of meat.

See Liao et al., 2012, p. 212.
3 See for example Gardiner, 2010; Hamilton, 2010.
4 Persson and Savulescu argue that moral bioenhancement does not make individuals unfree (Persson & Savulescu, 2012, pp. 112–115). For a critical discussion of this argument, see Harris, 2011.
5 For a productive overview, see Stehr, 2013.
6 Others may disagree, however. Some evolutionary anthropologists would argue that the whole point of institutions is to enable people to transcend their selfish interests. Indeed, institutions are what gave humans an evolutionary advantage.
7 See also Hardin, 1968 and Ophuls, 1977.
8 The reasons for downgrading France to the category of flawed democracy are (1) deterioration of media freedom, (2) extremely low public confidence in political parties and government, (3) declining engagement in politics, (4) low degree of popular support for democracy, (4) the widening gap between the people and political elites, (5) violent rioting as a symptom of the country's political malaise, (6) power concentration around the president, and (7) increased anti-Muslim sentiments. The reason for downgrading Italy to a flawed democracy is the politicised media. Greece has been downgraded because of low scores for government functioning and political culture. Furthermore, corruption in Greece has increased and government transparency and accountability are low. See Economist Intelligence Unit, 2011, pp. 16–17.
9 See United States Energy Information Administration, 2006; Held and Hervey, 2009, pp. 6–7.

References

Ansell, C. K. (2011). *Pragmatist Democracy: Evolutionary Learning as Public Philosophy.* Oxford: Oxford University Press.

Baber, W. F., & Bartlet, R. V. (2005). *Deliberative Environmental Politics. Democracy and Ecological Rationality.* Cambridge: MIT Press.

Beeson, M. (2010). The Coming of Environmental Authoritarianism. *Environmental Politics, 19*(2), 276–294.

Dryzek, J. S., & Stevenson, H. (2011). Global Democracy and Earth System Governance. *Ecological Economics, 70*(11), 1865–1874.

Economist Intelligence Unit (EIU) (2011). *Democracy Index, 2010: Democracy in Retreat. A report from the Economist Intelligence Unit.* London: Economist Intelligence Unit.

Economist Intelligence Unit (EIU) (2015). *Democracy Index, 2014: Democracy and its Discontents. A Report from the Economist Intelligence Unit.* London: Economist Intelligence Unit.

Frey, B. S., & Jegen, R. (2000). Motivation Crowding Theory: A Survey of Empirical Evidence. *CESifo Working Paper Series,* 245.

Fukuyama, F. (2005). *State-Building: Governance and World Order in the 21st Century.* Ithaca, New York: Cornell University Press.

Gardiner, S. M. (2010). Ethics and Global Climate Change. In *Climate Ethics: Essential Readings,* S. M. Gardiner, S. Caney, D. Jamieson & H. Shue (Eds). Oxford: Oxford University Press.

Fuller, S. (2012). The Future of Ideological Conflict. *Project Syndicate.* 7 May.

Hamilton, C. (2010). *Requiem for a Species.* London and Washington, DC: Earthscan.

Hardin, G. (1968). The Tragedy of the Commons. *Science, 162*(3859), 1243–1248.

Harris, J. (2011). Moral enhancement and freedom. *Bioethics, 25*(2), 102–111.

Hayek, F. (1973). *Law, Legislation and Liberty, Volume 1: Rules and Order.* Chicago: The University of Chicago Press.

Held, D., & Hervey, A. (2009). Democracy, Climate Change and Global Governance: Democratic Agency and the Policy Menu Ahead. *Policy Network Paper*, November. London: Policy Network.

Hurley, S. (2011). The Public Ecology of Responsibility. In *Responsibility and Distributive Justice*, C. Knight and Z. Stemplowska (Eds.), Oxford: Oxford University Press.

Jefferson, W., Douglas, T., Kahane, G., & Savulescu, J. (2014). Enhancement and Civic Virtue. *Social Theory and Practice, 40*(3), 499–527.

Lahsen, M. (2005). Technocracy, Democracy, and U.S. Climate Politics: The Need for Demarcations. *Science, Technology, & Human Values, 30*(1), 137–169.

Landemore, Hèlène (2012). Collective Wisdom: Old and New. In *Collective Wisdom: Principles and Mechanisms*, Hèlène Landemore and Jon Elster (Eds.), pp. 1–20. Cambridge: Cambridge University Press.

Le Grand, J. (2006). *Motivation, Agency, and Public Policy.* Oxford: Oxford University Press.

Liao, S., Matthew, A. S., & Roache, R. (2012). Human Engineering and Climate Change. *Ethics, Politics & Environment, 15*(2), 206–221.

List, C., & Goodin, R. E. (2001). Epistemic Democracy: Generalizing the Condorcet Jury Theorem. *Journal of Political Philosophy, 9*(3), pp. 277–306.

List, C., & Pettit, P. (2011). *Group Agency: The Possibility, Design, and Status of Corporate Agents.* Oxford: Oxford University Press.

Lovelock, J. (2010, March 29). James Lovelock on the Value of Sceptics and Why Copenhagen Was Doomed. Interview by Leo Hickman, *Guardian*.

Midlarsky, Manus I. (1998). Democracy and the Environment: An Empirical Assessment. *Journal of Peace Research*, Special Issue on Environmental Conflict. *35*(3), 341–361.

Olsen, J. (2007). *Europe in Search of Political Order: An Institutional Perspective on Unity/Diversity, Citizens/Their Helpers, Democratic Design/Historical Drift and the Co-existence of Orders.* Oxford: Oxford University Press.

Ophuls, W. (1977). *Ecology and the Politics of Scarcity, Prologue to a Political Theory of the Steady State.* San Francisco: W. H. Freeman.

Ostrom, E. (2005). *Understanding Institutional Diversity.* Princeton NJ: Princeton University Press.

Persson, I., & Savulescu, J. (2012). *Unfit for the Future? The Need for Moral Enhancement.* Oxford: Oxford University Press.

Popper, K. (2002). *The Poverty of Historicism.* Berkshire: Routledge.

Rayner, S., Kruger, T., Savulescu, J., Redgwell, C., & Pidgeon, N. (2009). The Oxford Principles. www.geoengineering.ox.ac.uk/oxford-principles/principles/

Rejesus, R. M., Mutuc-Hensley, M., Mitchell, P. D., Coble, K. H., & Knight, T. O. (2013). U.S. Agricultural Producer Perceptions of Climate Change. *Journal of Agricultural and Applied Economics, 45*(4), 701–718.

Robock, A. (2008). 20 Reasons Why Geoengineering May Be a Bad Idea. *Bulletin of the Atomic Scientists, 64*(2), 14–59.

Rostbøll, C. (2008). *Deliberative Freedom: Deliberative Democracy as Critical Theory.* Ithaca: State University of New York Press.

Scavenius, T. (2014). *Moral Responsibility for Climate Change. A Fact-Sensitive Political Theory.* Copenhagen: University of Copenhagen.

Scharpf, F. (1973). *Planung als Politischer Prozess: Aufsätze zur Theorie der Planenden Demokratie* [*Planning is a political process: a theory of the planning democracy*]. Frankfurt Am Main: Suhrkamp Verlag.

Searle, J. (2010). *Making the Social World: The Structure of Human Civilization.* New York: Oxford University Press.

Shearman, D., & Smith, J. W. (2007). *The Climate Change Challenge and the Failure of Democracy.* Westport, CT: Praeger.

Shue, H. (1988). Mediating duties. *Ethics, 98*(4), 687–704.

Stehr, N. (2013). An Inconvenient Democracy: Knowledge and Climate Change. *Society, 50*(1), 55–60.

Streeck, W. (2011). The Crises of Democratic Capitalism. *New Left Review, 71,* 5–30.

Sunstein, C. R. (2007). Deliberating Groups versus Prediction Markets (or Hayek's Challenge to Habermas). *Episteme, 3*(3), 192–213.

Surowiecki, J. (2004). *The Wisdom of Crowds.* New York: Doubleday.

United States Energy Information Administration (EIA) (2006). *International Energy Annual 2006.*

3 Institutional responsibility

Jessica Nihlén Fahlquist

Introduction

During the last 10 or 20 years, public debate in many countries has started to include questions of what causes climate change and how we should mitigate and adapt to it. These questions are not merely factual, but also normative. It is essentially an ethical question why we should solve climate change and who should do it (see Gardiner, 2004; Garvey, 2008). The discussion involves questions about moral responsibility. However, when discussed as a question of responsibility, it is crucial to make a distinction between backward-looking and forward-looking responsibility. Responsibility can either be backward-looking, based on notions of causation and blameworthiness, or forward-looking – focusing on, for example, solutions to a problem. When we discuss forward-looking responsibility for climate change, we inevitably discuss ability and capacity to respond.

However, we need to ask: what exactly is the link between being morally responsible for something and having the ability and capacity to act on that responsibility? This question is not adequately addressed in the philosophical literature on moral responsibility. The reason for this may be the traditional focus on backward-looking responsibility, i.e. being responsible for that which has already happened. Outside the realm of philosophy, it is common knowledge that responsibility can also be forward-looking, but this insight has traditionally not been adequately addressed in the philosophical debate, although there are exceptions (e.g. Richardson, 1999, pp. 218–249; Young, 2006; Goodin, 1986; Miller, 2005; Green, 2005; Nihlén Fahlquist, 2006b; Van de Poel, Nihlén Fahlquist, Doorn, Zwart, & Royakkers, 2012). In this chapter the question about responsibility for climate change is explored. The main ideas are the following:

1 Responsibility is linked to the options presented to the agent, and as such it is largely a matter of institutional design
2 Capacity entails forward-looking responsibility
3 A narrow focus on individual responsibility will not solve the problems associated with climate change.

Instead, because capacity and ability to act are closely linked to forward-looking responsibility, institutional actors are the primary subjects of responsibility in the case of climate change. These ideas are discussed in turn. However, it is useful to start analysis with a short introduction to the concept of moral responsibility.

Moral responsibility

The concept of moral responsibility is complex and used differently in different contexts. It is sometimes equated with blameworthiness. Blameworthiness has been the focus of philosophical analysis since Aristotle, but the term 'responsibility' is fairly new (McKeon, 1957).

In the context of climate change, responsibility is not merely a question about blameworthiness, but also about tasks and duties directed at future action-taking. The notion that responsibility is sometimes more forward-looking than backward-looking is common in non-philosophical discussions. Traditionally, the concept of responsibility analysed by philosophers is backward-looking; responsibility as blameworthiness or culpability. According to Strawson's influential theory of responsibility, moral responsibility consists of the reactive attitudes, for example resentment and gratitude, that we hold towards each other as co-members of the moral community (Strawson, 1962).

However, there is a difference between *being held* responsible and *being* responsible. In addition to reacting to others by holding them responsible, we can and regularly do analyse and criticise responsibility ascriptions, reflecting on whether they are reasonable or not. According to Fischer's revised Strawsonian theory, 'agents are morally responsible if and only if they are *appropriate* recipients of reactive attitudes' (Fischer & Ravizza, 1993). This notion indicates that we do sometimes excuse and exempt people when we do not think it would be reasonable to hold them responsible. Understandably, the philosophical discussion of moral responsibility is closely linked to the discourse of free will in which incompatibilists argue that free will is necessary for moral responsibility and compatibilists argue that even without free will, it makes sense to hold each other responsible.

Although the question of whether free will exists is relevant to environmental problems, the way we ascribe and distribute responsibility for societal problems like climate change is more practical. Leaving the philosophical discussion aside, there is also a social practice of ascribing responsibility, and arguably when responsibility is ascribed to an individual or collective agent for a societal problem, two concerns are important.

Responsibility ascriptions and distributions should ideally be both fair and effective (Nihlén Fahlquist, 2006b). They should be fair partly because it is, intuitively, wrong to hold someone responsible if, for example, they did not act voluntarily. Fairness is, of course, a highly contested concept. What is considered fair varies depending on, for example, normative theory and political ideology. In this context, it should simply be acknowledged that fairness is one of the values involved in responsibility ascriptions, but people will inevitably disagree on whether a particular responsibility ascription is fair and what fairness means.

When is it fair to hold someone responsible? Of course there is no simple answer to this question. In a sense, this discussion focuses on reasonable conditions for holding someone responsible in the sense of blameworthiness. In ordinary language as well as in the philosophical debate, a basic notion is that a person is responsible if certain conditions are fulfilled. For example, if someone was forced to do something that we normally consider wrong, we tend not to hold her responsible, but would excuse her for acting wrongfully.

The following is a useful list of conditions that are often seen as prerequisites for holding an agent morally responsible in the sense of backward-looking responsibility (Van de Poel et al., 2012; Nihlén Fahlquist, forthcoming):

- Capacity
- Causality
- Knowledge
- Freedom
- Wrong-doing

The first condition requires that the potentially responsible agent is a moral agent who has a certain capacity for rational thinking and moral deliberation. It is common to exempt some groups of human beings from responsibility, for example children and people with mental disorders, because they lack this capacity (e.g. Wallace, 1994; Van de Poel et al., 2012). Causation is a complex notion, but unless the agent is causally linked, in some way, to the action or consequence for which she might be held responsible, she may not be held responsible. Causation is sometimes fairly straightforward.

However, in some cases the causal links are less clear. For example, it has been argued that consumers in industrialised countries contribute to structural processes causing global injustice and that they are, therefore, responsible (Young, 2006). The way individuals contribute to climate change is another example of a relatively loose causal connection. However, it is clear that some, more or less tangible, causal connection is required for moral responsibility. Sometimes, when the term 'responsibility' is used, it is really the causes we are interested in. As we saw above, the conditions of knowledge and freedom have been seen as prerequisites of moral responsibility as blameworthiness since Aristotle. The most common excuses that people offer when they think they are wrongfully held responsible are 'I did not know' and 'I was forced'. The last condition requires that there has been some wrong-doing. What counts as wrong-doing differs depending on ethical theory, and possibly contextual considerations. However, we do not hold others responsible in the sense of blameworthiness unless there appears to have been some kind of wrongdoing (Van de Poel et al., 2012; Nihlén Fahlquist, forthcoming). These conditions are important when reflecting on the fairness of responsibility ascriptions.

When discussing real societal problems rather than mere philosophical or hypothetical cases, fairness is not the only relevant aspect. Responsibility ascriptions should also ideally be effective. To require that responsibility ascriptions are effective means that they should ideally contribute to solving societal problems. Whether it be public health, poverty, education or the environment – when distributing responsibility between different actors (e.g. individuals, governments, NGOs) fairness is not the sole concern. Responsibility should ideally be ascribed to the agent who is likely to solve the problem. This could be just as complicated as deciding on what is fair. There are epistemic gaps, and normative differences, as well as other obstacles to decisions concerning efficacy in responsibility distributions. However, most people would probably agree with the abstract notion that responsibility ascriptions should be both fair and effective, even if they might disagree on what that would mean in the case of, for example, climate change.

Responsibility is linked to options

Individuals act in ways that, on an aggregate level, cause climate change. They drive cars, go on holiday and eat meat. This fact has now been acknowledged by the media and in many countries has led to a discussion on individual responsibility for climate change. The idea emerged at the beginning of the 21st century that we can all make a difference in our everyday choices concerning, for example, transportation. This public image of climate responsibility goes hand-in-hand with the rationalistic understanding of agency that is being questioned in this book. The focus on individuals and their behaviour is not very effective in solving the problems we are confronted with, and it distracts our attention from the political and cultural context, i.e. the institutional frameworks within which individuals act and behave. As we have seen, voluntariness is closely linked to moral responsibility. The voluntariness with which individuals are presumed to act has been criticised in social science research (Jacobsen & Dulsrud, 2007). Arguably, voluntariness is not binary, but a matter of degrees, and different institutional designs yield different behaviours (e.g. Le Grand, 2003; Thaler & Sunstein, 2008). Individual behaviour and institutional design are linked, and this link should be acknowledged in the debate on climate change and responsibility for reasons of fairness as well as efficacy.

Voluntariness is linked to the options and capacity we have for action. It is reasonable, I would argue, to hold individuals responsible, but only to the extent that they have real options to act in ways that do not harm the environment. Unless an agent is presented with real options to act in environmentally friendly ways, she cannot be said to have acted completely voluntarily. Consequently, moral responsibility is also a matter of degree and needs to be contextualised. This means that different individuals are responsible to different degrees depending on context, options and ability. Policy choices concerning institutional design can affect voluntariness and, by implication, the degree of individual responsibility.

As we have seen, the idea that agents should be excused in circumstances when their actions were not fully voluntary has been asserted by philosophers since Aristotle (Aristotle, 2011). However, this reasonable idea appears not to be taken fully into account in the public debate on individual responsibility for environmental problems. When the public debate started to focus on individual responsibility the question was 'What are you doing, as an individual, that causes carbon emissions?' and 'How big is your carbon footprint?'; the assumption being that the amount of carbon emissions equals the amount of responsibility. It has not been adequately acknowledged that we all act in different contexts and have different degrees of voluntariness. Here are some examples to consider.

Imagine that in Agnes's residential area, which is socio-economically disadvantaged, there is no recycling station and Agnes, who cannot afford to buy a car, would have to use public transport to get to the nearest recycling station. Agnes is a single mother of small children, and it is in practice very challenging for her to take her waste to the recycling station. Her options are as follows:

1   Going to the recycling station and putting the cans in the appropriate container. In this case, she has three options:

 a   Bringing her three children, paying for bus tickets, and going to the recycling station.
 b   Hiring a babysitter, paying for a bus ticket and taking the bus to the recycling station.
 c   Borrowing someone's car to go to the recycling station.

2   Throwing the cans in the household trash.

If Agnes lived in a place where the recycling station was merely a short walk away or if she had someone who could help her by taking her waste to the recycling station for her or look after her children while she went there, that would have made a difference in terms of her options. Alternatively, if there was a smooth and inexpensive system to share a car with other residents in the local community, her options would be different. Her current options have been designed by politics, tradition, habit and culture. It is also imaginable that the options could be changed by new policies and decisions. Intuitively, it does not appear fair to hold Agnes fully responsible for choosing the second option, i.e. to throw the cans in the household trash.

A second example relates to the suggestion that consumers should choose 'climate smart' food over non-climate smart food. Imagine the consumer, Benjamin, who wants to reduce his carbon footprint. Benjamin, who lives in Sweden, is in his local grocery store with the intention to buy broccoli. He has two options:

1   Buy fresh locally produced broccoli.
2   Buy frozen broccoli produced in South America.

Benjamin has heard the experts talk about the value of buying locally produced food, wants to do the climate-smart thing, and chooses to buy the fresh broccoli from Sweden. Due to the cultivation and mode of production in South America, the frozen broccoli has been produced without using any fossil fuel. In the process of producing the Swedish broccoli, a considerable amount of diesel has been used. For this reason, the frozen broccoli actually has a smaller carbon footprint in spite of it being transported from South America (National Board of Trade, 2012). Should Benjamin be considered responsible for choosing the less environmentally friendly option? Many of us would be reluctant to hold him fully responsible because the causal links are not known to him and it is almost impossible to gather all the information at all times when buying consumer products and to always choose the right thing. Even if he wants to and truly tries to do the right thing, he fails. He could be considered causally responsible, but not blameworthy.

We have now looked at two cases in which the environmentally damaging act of an individual is put in the context of the options they had when acting. In the first case the problem is the practical challenges or the effort that has to be made in order to do the right thing (from an environmental standpoint). In the second example, the obstacle is epistemic, i.e. it is extremely difficult to know which is the right option. Hence, even if the individual wants to take responsibility it is extremely difficult to do that. In addition to practical and epistemic obstacles affecting the options, ability and voluntariness that individuals face in seeking to be environmentally friendly, we all know that different countries and regions have different infrastructures and legislation affecting the choices that individuals can be expected to make.

Furthermore, if climate-smart products are more expensive than environmentally damaging products, then socio-economically disadvantaged individuals will have more obstacles in behaving responsibly in relation to climate change. Finally, if we add other values besides climate-friendliness and sustainability, e.g. public health, it becomes even more complicated to make the right choices as an individual. For example, should we eat farmed salmon? Should we support local farmers in countries like Sweden by buying meat to support a varied agricultural landscape and local production or reduce our meat consumption to save the climate? Should we use nuclear power to save the climate or stop using nuclear power to support a sustainable society? These are extremely difficult questions, not only scientifically or scholarly, but even more so for individuals.

The focus on individual responsibility for climate change is not likely to lead to adequate measures in the case of climate change. It is not likely to be effective. However, as we have seen, it is not fair either. People have different resources and other contextual circumstances to deal with, which means that their abilities to act differ too. As we have seen, philosophers disagree on whether the truth of causal determinism would affect the reasonableness of holding others morally responsible. However, the intuition that we cannot be held responsible for something we could not avoid is strong. This notion is also a central, although debated, principle in philosophy and often ascribed to Kant, who allegedly argued that 'ought implies can', i.e. if an agent is obliged to do X she should also be able to do X (e.g. Montefiore, 1958). Whether we are in fact able or unable to do X is, in reality, not binary and obstacles to action could be physical, epistemological or practical (Baard, 2016).

Conceptually, if and when we hold individuals responsible for X the under-lying argument is that they contributed causally and are blameworthy for doing X. In this sense, *holding* individuals responsible presumes a backward-looking concept of responsibility. What has been said so far provides an argument against holding individuals responsible in this backward-looking sense, which essentially means blaming individuals when they do wrong. However, we have not yet touched upon the question of a potential forward-looking responsibility of individuals.

Capacity entails forward-looking responsibility

If an individual has all the financial, social and epistemic resources, i.e. has reasonable options to act in a climate-friendly way, it is reasonable to expect that she acts in such a way. If she acts in such a way, she can be said to take forward-looking responsibility. Forward-looking responsibility could primarily be con-ceived of in two ways: a) responsibility as a task or obligation or b) responsibility as a virtue or character trait. The underlying intuition is well described in a short sentence by Garvey (2008), discussing the claim that rich countries ought to do more than poor countries to combat climate change. He suggests that just as 'ought implies can', 'can implies ought' in some circumstances. Garvey does not elaborate this idea, but arguably there is something highly reasonable in his brief statement. One of the arguments for the principle 'Common but Differentiated Responsibilities' is based on a similar notion. Common but Differentiated Responsibilities is the principle stating that rich countries should bear a greater proportion of responsibility for climate change.

There are two very different justifications for the principle of Common but Differentiated Responsibilities. First, rich countries are said to have a greater responsibility to solve the problems of climate change because they, historically, contributed more to the emissions of carbon dioxide. Second, rich countries have a greater capacity primarily in terms of power and resources to solve these prob-lems. Whereas the former justification is in line with the backward-looking notion of responsibility, the latter is more in line with a forward-looking notion of responsibility. This is a reasonable idea applicable to individuals as well as to states. If we think about contributions that people make to charity organisations, in terms of both money and time, the reasonable underlying intuition is that we should expect more from someone who is wealthy and does not have to work than from someone who works full time for a modest salary.

If an individual is in a good position to do something about climate change she has a responsibility to do so. If an individual is less capable of doing so, she should be partly excused. There is no standard individual who has a standard share of responsibility. There are only particular individuals in particular socio-economic, cultural and political contexts.

Another example relates to the notion that rich countries have a responsibility to help poor countries in various ways, with money, education and technology transfer. In developed, democratic, and well-ordered states, it is a widely shared notion that the government should give aid to less developed countries. There is often disagreement on the amounts of aid and the exact content (money or self-help aid), but most people agree that there ought to be some kind of aid to people in developing countries.

These are examples illustrating a widely shared, and I think reasonable, intuition that if an agent, whether individual or institutional, has the capacity, power, and resources to contribute to solving a societal problem, they have a responsibility to do so. This intuition underlies the so-called Rescue Doctrine, which states that if the act of a negligent person A causes wrong to B and C intervenes, A is liable for any danger C is exposed to in the process. The idea is that 'danger invites rescue' (Legal Definitions, 2017). This kind of responsibility is forward-looking, i.e. it does not hinge on the agent's causal contribution to the problem at hand. In addition, it is fairly open-ended and it does not have to be stated exactly when such a responsibility has been fulfilled. Consequently, it leaves open exactly what the agent ought to do and involves a certain degree of flexibility and permission to improvise (see Goodin, 1986).

This is the way in which individuals as consumers and citizens are responsible for climate change. Individuals should do what they can against the background of their capacity. The more resources, power and capacity an agent has the better her ability to contribute to solving the problem and the more reasonable it is to ascribe forward-looking responsibility to her. Resources includes several factors, e.g. money, but also education, information and leadership skills in terms of a talent for influencing others.

What, then, can be said about the reasonableness of alternatives? When does an individual consumer or citizen have reasonable options? The idea of reasonable alternatives can be stated in similar terms to the notion of voluntariness as a matter of degrees. Although consumers are not coerced into buying food that was produced using unnecessarily high levels of energy, if climate-smart food is very expensive compared to regular food, the alternative of buying the environmentally less damaging food is not really an option unless the consumer is wealthy. This is an example of how the cost of choosing the 'right' option is highly relevant. In addition to cost, the availability of good options is relevant. If 99% of the products in grocery stores are not climate-smart products, this is obviously an obstacle to people who want to reduce their carbon footprint even if they can afford to buy those products.

Similarly, unless there are safe and extensive bicycle lanes, the alternative to ride a bike instead of driving a car is not a realistic option for people who drive to work. The latter is an example of how culture, tradition and political decisions affect how people choose to act. They may choose to drive their car because they grew up and live in a society that treats cars and highways as very important parts of life and society. On the other hand, they may resist that cultural pressure and want to ride a bicycle or use public transport instead, but if the infrastructure makes it difficult or too inconvenient to choose those options, the conscientious citizen may, very reluctantly, continue to drive their car to work. Finally, there are complicated conflicts of values that could be stated as a normative obstacle to individuals. For example, how should an individual prioritise between health and environment, as the example of farmed salmon illustrates?

In real life, compulsion appears to be a matter of degree and although nobody is forcing people, it may be very difficult to choose the environmentally friendly option. This notion has not been fully acknowledged in public debate or the scholarly analysis of climate change embracing the rationalistic understanding of individual agents.

It is obviously difficult to draw a line between what is reasonable and what is not in individual cases. On the political and general level, it is clear that some infrastructural, cultural, socio-economic and political features of modern societies are questionable from an environmental perspective and that these features affect to what extent individuals are able to promote solutions to environmental problems. The most obvious problem is likely to be the car-dependency and meat consumption of industrialised nations. These problems are based on culture and tradition, but also lack of political initiatives and leadership.

In addition to cost and the availability of good options, there are epistemic obstacles. Information about the environmental footprints we leave is obviously crucial. However, the question of what information is available is often complex. In many cases, the information does not reach all groups of people in society. Furthermore, impeding factors, e.g. cost and availability, affect the reasonableness of the options that individual consumers and citizens have when it comes to acting in environmentally friendly or environmentally unfriendly ways. For these reasons, it is neither fair nor effective to ascribe backward-looking responsibility to individuals.

Institutional responsibility

The argument above can be summarised as follows. It is, generally, not fair to ascribe responsibility in the backward-looking sense, i.e. to blame individuals for their individual carbon emissions unless they have reasonable alternatives and resources to act in environmentally friendly ways. It matters, from a moral point of view, whether an individual has access to, for example, recycling stations and public transport. However, it is fair to ascribe forward-looking responsibility to individuals, based on their capacity to contribute to solutions to environmental problems. Furthermore, a considerable share of forward-looking responsibility should be ascribed to institutions because they can make the group of capable, hence responsible, individuals larger through policy-making and institutional design. We will now turn our attention to institutional responsibility.

If we consider Agnes and Benjamin above, institutional design could change the nature and number of options, at least to some extent. If ability, resources and options are crucial to forward-looking responsibility, the question is who is best able to respond to the challenges presented by climate change. In essence, who has the capacity to do something about this gigantic challenge?

Although it is beneficial to the issue at hand that most individuals in the developed part of the world are now aware that what we do as human beings affects the climate, there is a risk that too much focus is put on individual behaviour, at the expense of illuminating the vital role of institutions. If we agree that the options individuals have and that the contexts and abilities are different, the question arises how the context can be made to facilitate climate-smart behaviour. The underlying question is: who has the capacity to effectively affect the aggregate negative impact on climate change? This is where we need to start discussing the institutions. Institutions are important not only from an economic or political perspective, but from an ethical perspective. They are important because of the link between capacity, options and moral responsibility. The essence of my ethical argument for institutional responsibility is the following. Institutions have a responsibility to create options and design contexts in which individuals have a practical, economical and epistemic opportunity to behave responsibly in relation to climate change. Ascribing responsibility to institutions is motivated primarily by the efficacy aim of responsibility ascriptions. Simply put, if we ascribe responsibility to institutional actors we have a better chance of creating a society in which there are real opportunities to act in an environmentally friendly way.

There is an interesting philosophical discussion on the ontology of collective entities and whether they are no more than the sum of the partaking individuals. However, this chapter is not engaged with the ontology, but adopts a more commonsense or layman's approach to collective agents. Institutional, i.e. collective, agents, here refers primarily to governments and corporations, but also mass media, local communities and NGOs. As described by Green, institutional agents are 'organized agencies', i.e. 'entities that can make decisions and act on them' (Green, 2005).

Thus, institutional agents are responsible not only because they are more powerful and can do more, but also because it is in their power to create reasonable alternatives for individuals. They have it in their power to make it easier and less expensive for individuals to choose the environmentally friendly option and they can provide information that is easily accessible and as straightforward as possible. In essence, they are responsible because they have the power to create opportunities for individuals to do what is right. Another way of phrasing it is to say that institutions can make it easier for individuals to assume forward-looking responsibility.

This could be done by making information easily accessible, subsidising climate-friendly food while taxing climate-damaging food, by product development and presentation of products and so forth. These would be minimal requirements. Additionally, the problems of climate change mitigation and adaptation need to be discussed and deliberated in society. The greater the extent to which institutional agents have taken their forward-looking responsibility, the greater the extent to which it is reasonable to ascribe both backward-looking and forward-looking responsibility to individuals when they do not choose the environmentally friendly option. First, the greater the availability and affordability of good options, the more reasonable it is to blame those individuals who still do not adjust their behaviour. Second, the greater the extent to which parties such as governments and corporations, media actors, local communities and NGOs and local communities have assumed their responsibility, the larger the group of individuals with enough capacity and resources to assume their forward-looking responsibility.

For example, if the government has invested in an extensive public transport system, the degree of individual responsibility for choosing to drive a car instead of using public transport is greater than it would have been if the public transport system had been underdeveloped and unreliable. If a corporation can provide environmentally friendly products at a reasonable cost it is their responsibility to do so.

One might argue that individuals have a personal responsibility, which would be eroded if we allocate too much responsibility to institutional actors. Do individuals not have to do anything by themselves? I think there are two ways to respond to such criticism.

First, one way of conceptualising a forward-looking responsibility is through a virtue-ethical approach. Williams views responsibility as a virtue, which essentially represents a 'readiness to respond to a plurality of normative demands' (Williams, 2008) This is a slightly different way to conceptualise the idea that individuals are too complex to assess morally merely on the basis of isolated actions. Instead, the focus should be on an individual's whole life and character as well as the way in which the character evolves and improves. It focuses on the different roles an individual may have and the challenges they face in responding to a plurality of sometimes conflicting demands. Viewed from that angle, individual responsibility is still very important. However, it should be acknowledged that individuals' character and the virtue of behaving responsibly can be affected by social systems, policies, and information and education. Against this background, institutional responsibility can be stated as follows: it is the responsibility of institutions to create systems that make it easier for individuals to respond to the emerging reasonable norm that we all ought to act in environmentally friendly ways.

A second way of responding to the critique that personal responsibility is eroded if too much responsibility is ascribed to institutions is the following. As discussed above, ascribing and distributing responsibility is a social practice that serves two purposes. First, it creates or establishes fairness. Second it is a tool to establish an effective division of labour in order to solve societal problems. The optimal distribution of responsibility is both fair and effective, although it is sometimes difficult to achieve both to the same extent. Sometimes, one of the two purposes is more important than the other and the two have to be weighed against each other case-by-case.

Arguably, the long-term goal should be to encourage people to cultivate green virtues (see Jamieson). However, from the short-term perspective we need to add that this distribution of responsibility should also be effective, i.e. contributing to a solution to the problem. That is why the greatest share of responsibility for environmental problems should be ascribed to the most powerful, resourceful, and capable actors, i.e. governments and corporations, media actors, local communities and NGOs, because they can create systems and norms that make it easier and less costly for people to choose the environmentally friendly option than to choose the environmentally harmful option. As argued by Shue (1998), some duties should be assigned to institutions instead of individuals because that is likely to be more efficient. Institutions can facilitate the coordination and cooperation that are needed for those duties to be fulfilled. A second reason is that it would be to demand too much of people to assign such duties to individuals because individuals have rights as well as duties and should be allowed some time outside of their role as duty-bearers. Thus, viewing institutions as the main duty-bearers is more effective, but also more fair than treating individuals as the main duty-bearers in the case of climate change. However, this does not mean that individuals are completely exempted. On the contrary, it is their duty to make sure there are adequate institutions to implement the duties in question.

Whereas Shue argues that the role of institutions is to implement the duties, Green (2005) argues that the responsibility of institutions is greater than the responsibility of individuals. He argues that while it is reasonable to keep the restrictive version of responsibility, i.e. the responsibility that always traces the individuals' harmful behaviour, a more comprehensive kind of responsibility should be assigned to institutions because they constitute a different kind of agent. Institutions have more power and can alter mass behaviour, they are better at collecting and processing information about direct and indirect consequences of their actions and they can spread the cost through taxation. Essentially, institutional agents have more capacity; hence a greater share of responsibility is justified. Similarly, Walter Sinnott-Armstrong (2005) argues that whereas individuals do not have a moral obligation not to waste gas, governments have a moral obligation to fight global warming, primarily due to the scale of the problem. Individuals inevitably have different roles, as consumers, citizens, parents, etc, and take part in different institutional arrangements They share, develop and question norms and rules and take part consciously or not in maintaining or changing the institutions they take part in. Consequently, it appears that individuals shape the behaviour of institutions and institutions shape the behaviour of individuals. For this reason, it may not be as easy as one may have thought to distinguish between individuals and institutions, which makes the question of individual and collective responsibility highly complex.

Even though it is hard to make a sharp distinction, due to the urgency and scale of climate change, it appears that a good case can be made to include institutions in the discussion on how to distribute responsibility for such problems. It appears reasonable to expect individuals with capacity, resources and knowledge to take a great share of responsibility for creating and maintaining environmentally friendly institutions. As argued by Sinnott- Armstrong, instead of just withdrawing from society and adjusting one's own lifestyle to create as little environmental damage as possible, it is even more important to be proactive and work to change government policies and laws (Sinnott-Armstrong, 2005).

Conclusion

In this chapter, I have discussed the concept of moral responsibility in relation to the problem of climate change. The notion of the standard rational individual who is fully responsible for climate change was criticised. Instead, it was argued that individuals should only be held responsible to the extent that they have reasonable alternatives. Today, many individuals lack such options or do not have the resources to contribute to a solution to climate change. There are socio-economic, epistemic and cultural obstacles to climate-friendly behaviour. Instead, we should turn our attention to institutional responsibility and ways in which institutional design can facilitate responsible behaviour. However, just as institutions affect the behaviour of individuals, individuals shape the behaviour of institutions. Individuals and institutions are inevitably intertwined. Thus, the only reasonable notion of responsibility for climate change acknowledges the complexity and the need to move beyond the simplistic notion of individual responsibility.

References

Aristotle (2011). *Nicomachean Ethics*. Edited by Crisp, R. Cambridge: Cambridge University Press.

Baard, P. (2016). Risk-reducing goals: ideals and abilities when managing complex environmental risks. *Journal of Risk Research, 19*(2), 164–180.

Brewer, J., & Trentmann, F. (Eds.) (2006). *Consuming cultures, global perspectives: Historical trajectories, transnational changes*. Oxford: Berg.

Fischer, J. M., & Ravizza, M. (1993). Introduction. In J. M. Fischer & M. Ravizza (Eds.), *Perspectives on moral responsibility*. Ithaca: Cornell University Press.

Gardiner, S. (2004). Ethics and global climate change. *Ethics, 114*(3), 555–600.

Garvey, J. (2008). *The ethics of climate change. Right and wrong in a warming world*. London: Continuum.

Goodin, R. E. (1986). Responsibilities. *Philosophical Quarterly, 36*, 50–56.

Green, M. (2005). Institutional responsibility for moral problems. In A. Kuper (Ed.), *Global responsibilities who must deliver on human rights?* New York: Routledge.

Jacobsen, E., & Dulsrud, A. (2007). Will consumers save the world? the framing of political consumerism. *Journal of Agricultural and Environmental Ethics, 20*, 469–482.

Jamieson, D. (2007). When utilitarians should be virtue theorists. *Utilitas, 19*(2), 160–183.

Ladd, J. (1991). Bhopal: An essay on moral responsibility and civic virtue. *Journal of Social Philosophy, 32*(1), 73–91.

Legal Definitions. US Legal, definitions.uslegal.com/r/rescue-doctrine/ Accessed 3 May 2017.

Le Grand, J. (2003). *Motivation, Agency, and Public Policy: Of Knights and Knaves, Pawns and Queens*. New York: Oxford University Press.

McKeon, R. (1957). The development and the significance of the concept of responsibility. *Revue Internationale de Philosophie, XI*(39), 3–32.

Micheletti, M. (2003). *Political virtue and shopping: Individuals, consumerism, and collective action*. New York: Palgrave Macmillan.

Miller, D. (2005). Distributing responsibilities. In A. Kuper (Ed.), *Global responsibilities. Who must deliver on human rights?* New York: Routledge.

Montefiore, A. (1958). 'Ought' and 'can' *The Philosophical Quarterly, 8*(30), 24–40.

Nihlén Fahlquist, J. (2006a). Responsibility ascriptions and vision zero. *Accident Analysis and Prevention, 38*, 1113–1118.

Nihlén Fahlquist, J. (2006b). Responsibility ascriptions and public health problems. Who is responsible for obesity and lung cancer? *Journal of Public Health, 14*, 15–19.

Nihlén Fahlquist, J. (2009). Moral responsibility for environmental problems – individual or institutional? *Journal of Agricultural and Environmental Ethics, 22*(2), 109–124.

Nihlén Fahlquist, J. (Forthcoming). Responsibility analysis. In Sven Ove Hansson (Ed.) *Methods or the Ethics of Technology*. Rowman and Littlefield.

Paul, E. F., Miller, F. D., & Paul, J. (Eds.) (1999). *Responsibility*. Cambridge: Cambridge University Press. Special Issue National case studies of institutional capabilities to implement greenhouse gas reductions.

Richardson, H. S. (1999). Institutionally divided moral responsibility. In E. F. Paul, F. D. Miller & J. Paul (Eds.), *Responsibility*. Cambridge: Cambridge University Press.

Sassatelli, R. (2006). Virtue, responsibility and consumer choice: Framing critical consumerism. In J. Brewer & F. Trentmann (Eds.), *Consuming cultures, global perspectives: Historical trajectories transnational changes* Oxford: Berg.

Shue, H. (1988). Mediating duties. *Ethics, 98*(4), 687–704.

Sinnott-Armstrong, W. (2005). It's not my fault. In W. Sinnott-Armstrong & R. B. Howarth (Eds.), *Perspectives on climate change: Science, economics, politics, ethics*. Amsterdam: Elsevier.

Strawson, P. (1962). Freedom and Resentment. In Proceedings of the British Academy 48. (Reprinted from Perspectives on moral responsibility, J. M. Fischer & M. Ravizza (Eds.), 1993, Ithaca: Cornell University Press.

Thaler, R., & Sunstein, C. (2008). *Nudge. Improving Decisions about Health, Wealth, and Happiness*. New Haven, CT: Yale University Press.

Van de Poel, I., Nihlén Fahlquist, J., Doorn, N., Zwart S., & Royakkers, L. (2012). The Problem of many hands: climate change as an example. *Science and Engineering Ethics, 18*(1), 49–67.

Wallace, R. J. (1994). *Responsibility and the moral sentiments*. Cambridge; MA: Harvard University Press.

Williams, G. (2008). Responsibility as a virtue. *Ethical Theory and Moral Practice, 11*(4), 455–470.

Young, I. M. (2006). Responsibility and global justice: A social connection model. *Social Philosophy and Policy, 23*(1), 102–130.

Part II

Response capacity and behavioural change

4   The double gap between climate values and action

Theresa Scavenius and Malene Rudolf Lindberg

Introduction

The current dominant social scientific approach to the study of climate action is driven by an economic bias (Bjurström & Polk, 2011). Many climate studies focus on individual agency and economic incentives. This behavioural trend in the scientific realm corresponds to the economic-incentive focus in the political realm. The political debates on climate action focus on citizens' incentives to act and the willingness to pay for green solutions. The economic bias in scientific and political exchanges has the unfortunate consequence of marginalising a central aspect of the political climate challenge. Because the focus of the economic bias on individual actors' behaviours and incentives overestimates the institutional and political conditions that enable or hinder individual behaviours, it removes a large section of climate political insights and potential strategies from the discussion.

We argue in this chapter that the climate challenge requires a break with the focus on individual actors' behaviour and choices. The chapter suggests a 'context-sensitive' approach to behavioural studies that contributes to the conversation about the importance of developing societal climate resilience and institutional capacity. The context-sensitive approach builds on a long tradition in the sociological and anthropological literature that focuses on the societal structures, institutions and norms that frame and contextualise individual agency (Rayner, 1991). The arguments that we pose contribute to a growing literature on the non-attitudinal factors that approach climate actions and failures of climate action that is different from the current dominant climate scientific paradigm characterised by the economic bias and a focus on the individual level of agency (Bjurström & Polk, 2011). Our argumentation and conclusions in this chapter are based on an empirical survey conducted in Denmark in 2015. The most interesting finding during our work with the survey data was indications of what we call a *double gap in climate action*. The double gap points to the contradiction that even though pro-environmental attitudes and incentives increase pro-environmental action, many green attitudes are not translated into climate action, while on the other hand, much of the climate action undertaken by individuals in Denmark seems to be guided by contextual factors rather than by individual incentives. This double gap between individual incentives and actions problematises a behaviour and incentives solution to the climate challenge and explains why incentive-oriented politics has caused a vast climate action deficit.

Behaviour and incentive based approaches

For the last three decades, the main political goal in climate politics has been to motivate behavioural changes with economic incentives (Scavenius, 2017). A good example of this type of climate politics is the 1997 Kyoto Protocol, which established a global market for CO_2 quotas. At the national level, climate politics is also monetarised and individualised in the sense that the main climate political concern is about people's economic incentive to change light bulbs, petrol-based vehicles, etc. instead of large infrastructural investments in green technologies and energy resources.

These climate policy strategies today revolve around intervention in individuals' behavioural patterns through persuasion, pricing or advising (Clapp & Thistlethwaite, 2012; Shove, 2010). The provision of incentives is usually organised within market institutions that are believed to be the most effective governance structure. Climate politics today predominantly use economic and rational-instrumental governing models in contrast to political regulation and bans (Aldred, 2016, p. 148).

The focus on individual CO_2 emissions is prompted by an overlapping interest between several branches of climate science and climate discussions of moral responsibility, global inequality and justice. Climate science, climate philosophy and climate economics converge in an emissions-based and individual-oriented understanding of climate challenges (Scavenius, 2014; Scavenius & Rayner, Introduction, this volume).

An unfortunate result of this alliance between science, philosophy and economics is that the political framework they agree on is blind to the social and political structures that condition human behaviour. The dominant narrative of climate politics marginalises the analytical focus on institutions, infrastructure and social and political contexts (Sagoff, 2008; Shove, 2010). Bibliometric studies demonstrate an economic bias in social scientific studies of climate change (Bjurström & Polk, 2011). The consequence of this economic thinking is that both political and scientific attention is directed to an extreme degree toward individual actors' micro-actions. This focus neglects the importance of the societal context within which individual agency occurs.

The failures of the ABC model

Shove has famously labelled the rational paradigm of behaviour the 'ABC' model, which means 'attitudes' cause 'behaviour' and 'choice' (Shove, 2010). The ABC model fits the current framing of climate change as a challenge at the level of individual behaviour and individual responsibility (Shove, 2010, p. 1274; Scavenius, unpublished). Policies are developed to engage people's moral inclination when they can choose between products that are CO_2-neutral or not, and when they can offset their climate footprint when buying a bus ticket in Germany.

Many scholars, including Shove and Hurley, have criticised the rational assumptions underlying the ABC model. It presumes that people's behaviour will change if they receive more knowledge and thus gain awareness of and understand the incentives for green choices. For example, if an individual agent is presented with new knowledge on climate change or a new policy strategy to incentivise pro-environmental acts, the agent will change his or her attitude towards the environmental issues accordingly, which in turn will cause more pro-environmental behaviour. There is obviously some truth is this. When it is possible to offset your climate footprint when riding a German bus, the attention of bus passengers turns to the climate issue, and when at the same time passengers are provided with an economic compensatory action option, many people will choose to offset their footprints. But this does not mean that the rationalistic framing of agency is the most productive approach to environmental and climate politics.

A relevant question is to what extent social scientific studies of environmental and climate politics manage to go beyond the behaviour-change initiatives studies. As Shove argues, the ABC paradigm in environmental politics has marginalised much relevant social theory by assuming a behaviouristic approach to attitudes and choices: individual agents are solipsistic and can develop their attitudes independently of social and political context. And individual agents are capable of voluntarily choosing their behaviour.

Many political science studies take this economic approach to climate behavioural studies. Often this is done in surveys in which individuals are asked about how much they are willing to pay for climate initiatives or how much more they are willing to pay for climate-friendly products compared to other types of products (Eurobarometer, 2008). Whereas these survey studies provide information about people's self-reported attitudes towards price and climate-friendly products, they say very little about the institutional, social and political barriers and contexts of climate behaviour. The ABC model obscures the extent to which individuals' choices are affected by structural factors in society and also neglects the regulatory role of collective actors such as the state, businesses and producers.

As our empirical findings reveal below, we find that economic incentives and persuasion, and pricing and advice, matter to Danes. However, our findings also indicate that there are other explanatory forces that influence *why* people choose to act or *not* to act that are at stake here. We find that respondents point to non-individual, collective and contextual factors as determining factors of their (lack of) climate behaviour. This might indeed be a comfortable excuse for non-action, but it might also point to the importance of paying less attention to individual actors and more attention to collective actors, both in climate studies and climate policies. Collective climate action deals with technology, resource management, effective supply and capacity building, and these do indeed seem important in determining the climate footprints of individuals.

The economic behavioural studies black-box contextual factors about cultural identity and institutional lock-in problems, presuming that people often do not have alternative options with regard to transportation and energy usage. Below, we report findings from a Danish survey that reveal interesting insights into how the challenges of individual action-taking while embedded in the specific institutional setting of Denmark are experienced. We were very careful to design the survey so that the questionnaire did not pre-cast economic assumptions about willingness to pay. In contrast, we asked people about their access to climate-friendly transformation, energy systems and infrastructure.

The double gap

The survey was conducted in Denmark in 2015 and was designed to investigate individuals' perceptions of facilitators and barriers to environmental and climate behaviour. The most interesting finding during our work with the survey data was indications of what we call a *double gap in climate action.* The double gap points to the contradiction that even though there is a significant correlation between environmental attitude and environmental behaviour, green values are not necessarily translated into green action; while on the other hand green actions are not necessarily guided by factors other than green values.

Using data from this survey of a representative sample of the Danish population,[1] we examined the correlation between green values and green actions in Denmark. We presented participants with eight different green action opportunities requiring different amounts of energy, time and money. These opportunities vary from energy saving and waste sorting at home, environment-conscious consumption of groceries and consumer goods, and green transportation choices, to conducting climate renovations and weighing environmental considerations when buying a car or a home or travelling. Informants were also asked about climate attitudes and green values and how important they find the climate issue compared to other political questions. Among other things, informants were asked to state the degree to which they see themselves as a person with green values on a scale from 1–10. These statements were used as indicators of a person's climate *attitude*. The total amount of green actions undertaken during the last six months was used as indicator of individual's climate *action* level. The maximum action score possible was 8.

On the basis of the ABC model of behaviour, we would expect two things: 1) people with a pro-environmental attitude, expressed here as a high self-reported green value score, would conduct many green actions, as the green attitude would translate into green choices and behaviours. We would also expect that 2) people with a low green value score would conduct few or no green actions, as their lack of environmental attitude would be a barrier to green choices and behaviours. One might expect the relationship between green attitudes and green actions to look somewhat like the line illustrated in Figure 4.1.

Here, people with a very low value score are not expected to conduct climate actions, as they have no or little incentive to do so. Thus, in Figure 4.1, we are

Figure 4.1 Expected relationship between green actions and green values.

not expecting people with a self-reported green value to score under 3.5 for their conduct of green actions. As value score increases, people are expected to undertake a growing number of actions. People with the maximum green value score are expected to conduct the maximum amount of action.

However, the observed relationship between climate values and actions looks somewhat different. Figure 4.2 illustrates the relationship between the green actions and green values reflected in the survey.

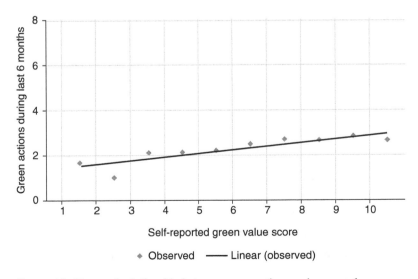

Figure 4.2 Observed relationship between green actions and green values.

The survey data reveal a significant positive linear relationship between value and action, which is shown by the black line in the graph. This supports the idea of a positive correlation between climate attitude and climate action: the higher value score, the more green actions undertaken ($p=0.00$).

However, we were surprised by the nature of the relationship between values and actions identified. Following the rational-incentive assumptions in much scientific and political work, we would expect people with very low value score to have very little incentive to act greenly, and thus not conduct any actions. Nevertheless, this was not the case. Even people with a value score of 5 or less conduct an average of 1.1 out of the 8 possible climate actions ($n=350$). By comparison, people with a value score above 5 conducted an average of 2.5 climate actions out of the 8 possible ($n=656$). Even though there is a significant positive linear relationship between climate attitude and action in Denmark, it does not ascend to the degree we expected given the ABC model's assumption of correspondence between attitudes and behaviours. Thus, the observed relationship differs from the expected relationship, as the slopes of the lines in the two graphs differ significantly, indicating a stronger expected relationship than we were able to identify empirically. The line in Figure 4.2 is far from as steep as we would expect given the ABC model's understanding of behaviour.

The graph in Figure 4.3 combines the expected and observed relationship between climate action and attitude to illustrate the discrepancy. Two gaps between the expected and observed relationships are revealed.

First, people holding many green values are not conducting a corresponding number of green actions. This is illustrated as GAP I. The discrepancy denotes a value–action gap parallel to gaps identified in previous studies conducted in the

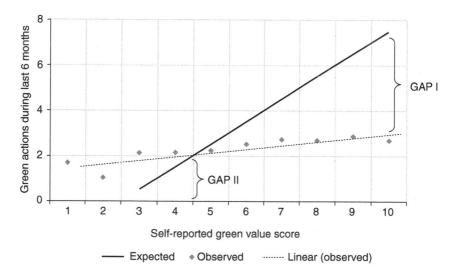

Figure 4.3 Relationship between green actions and green values.

UK (see e.g. Blake, 1999; Flynn, Bellaby, & Ricci, 2009). Blake (1999) was the first to point out such a gap and to call it *the action–value gap.*

Second, people with few green values conduct more green actions than expected. This is illustrated as GAP II. In spite of low levels of climate-friendly attitudes, people in Denmark conduct some green actions nonetheless. This second gap indicates that the value–action gap is allegedly a double gap: not only do Danes with strong climate-friendly attitudes conduct fewer green actions than expected; Danes with no or weak climate-friendly attitudes also conduct more green actions than expected on the basis of the rationalistic approach in climate science and politics.

When it comes to climate issues, we identify how Danes tend to act in ways that do not correspond to their values. Based on the survey data from our study investigating attitudes and actions of Danes, two conclusions can be made: first, that people with green values act less greenly than the ABC model would call for, which points to the existence of a *value–action* gap in Denmark corresponding to similar gaps identified in other European studies of climate behaviour (e.g. Blake, 1999). Second, people with few green values act almost as green as people with more green values, which is indicative of a reverse value–action gap. We call this the *double gap:* it looks as if Danes tend towards exhibiting similar levels of green actions and that values only affect their behaviour in minor, albeit significant, ways. However, neither very climate-aware Danes or less climate-aware Danes are exhibiting green behaviours similar to what we would expect on the basis of the ABC model.

As a continuation of this line of thought, we can return to the two assumptions we made on basis of the ABC models, namely that: 1) people with a high self-reported green value score would conduct many green actions, and 2) people with a low green value score would conduct few or none green actions. Neither of these two expectations is satisfied in the survey results. There is a huge gap between the expected and observed relationship between values and actions in the Denmark study, which indicates a weaker relationship between green values and actions than the market models presume. Even informants who scored themselves relatively low on the green value scale conduct green actions. These acts cannot be accounted for in the causal behavioural models, as behaviour not driven by values and attitudes is inexplicable.

Contrary to the assumption that people act according to their beliefs, we see how the Danes, almost independently of their level of green attitude, have adopted green behaviour to some degree. Even people with few or no green values conduct one or more green actions. This points to the existence of influencers on environmental behaviour beyond the individual level. To understand the behaviour of Danes, attention must be given to the context surrounding individuals.

Our study offers indications that the value–action gap is a dual phenomenon: people hold green values that do not translate into green actions because of contextual restrictions. Simultaneously, green actions are conducted by people who are motivated by factors other than a wish to be environmentally friendly. This suggests that contextual factors function not just as barriers to green action, but also as accommodators of green action. Just as the absence of a well-functioning public transport system can be a hindrance to green transportation choices, the existence of such a system can promote climate-friendly transportation habits among people who do not hold green values. On the basis of our study, we conclude that individuals are not choosing and acting in a societal vacuum. Every day, people exhibit behavioural patterns that are inflexible and dependent on societal surroundings and institutional infrastructure.

Locked-in behaviours

As Sanne (2002) has argued, individuals are essentially 'locked-in' to societal configurations. The double-gap finding can be interpreted as evidence of this locked-in tendency. Here, our findings are in line with other findings (see e.g. Urry, 2008). Focusing on mobility and travel, Urry stresses how people are fundamentally locked-in to certain climate-damaging patterns of behaviour. The double gap finding shows that economic-structural factors may both hinder and encourage climate-friendly behaviours independently of individual attitude. Thus, understanding climate action within a framework of individual rational environmental behaviour leads both scientific work and politics to ignore important behaviour-shaping contextual factors. For example, many environmentally significant actions are highly constrained by infrastructure and institutional opportunities, which render personal values, beliefs and attitudes less significant with regard to behavioural outcome. These institutional opportunities point to the need for climate policies and social change that do not focus on changes in individual attitudes and values.

Individual actors are often locked-in to action patterns determined by political, social and infrastructural circumstances (Urry, 2008). Climate-friendly actions require a behavioural space; this behavioural space is dependent on political legislation, investments and regulations. For example, a transition from private to public transportation requires a large and stable public transportation network. Likewise, the separation of waste is important at the individual level only if the necessary infrastructure and waste management systems are established to deal with segregated waste streams. Without an action-supporting context, citizens are locked-in to climate-sensitive behavioural patterns.

Complex incentive structures

The double-sided lack of correspondence between green *values* and green *actions* found in our study fundamentally challenges any decision based on the assumption of the ABC framework. By arguing that the ABC model is an insufficient method to understand climate behaviour and to create solutions for the current societal challenges, our study is in line with the path that the scientific literature on climate behaviour has developed.

Since the rise of scholarly interest in environmental behaviour in the early 1970s, we have witnessed an evolution in the field from focusing solely on rational agents behaving according to attitudes to also focusing on an increasing number of social contextual factors that constitute various features of social, economic and political surroundings (Kollmuss & Agyemann, 2002; Stern, 2000; Flynn et al., 2009; Gausset, Hoff, & Scheele, 2015). In contrast to the economic models, these include non-attitudinal factors.

Accumulated research in climate behaviour shows that reasons for green actions are varied and incentive structures are complex. Since the 1980s, studies have found that climate and environmental behaviour is significantly related to a number of non-economic factors, including demographic, individual and contextual ones. Important demographic factors are gender and level of education (Gundelach et al., 2012). Individual factors include knowledge about climate and environmental conditions (Thøgersen, 2005), values and norms (Ajzen & Fishbein, 1980), emotions (Antonetti & Maklan, 2014), ethical and moral beliefs (Stern, 2000) and degree of sense of responsibility (Blake, 1999). The context within which individuals act seems also to shape the actions of individuals. For example, research has found that culture, legislation and political context (Stern, 2000), social networks (Gausset et al., 2015) and available infrastructure and institutional context (Flynn et al., 2010) have a decisive influence on individuals' climate behaviour. Furthermore, in Denmark, we see that health gains and concerns for the next generations are important motivational factors for green action (Gausset & Hoff, 2015). Individuals' incentive structures are thus more complex and more closely intertwined with the social and political context than the ABC framework recognises. Our empirical study in Denmark contributes to the scope of scientific literature that argues that individuals' incentive structures are more complex than standard models of individual green attitudes and actions would have us believe.

Political implications

Our study points to the importance of a 'context-sensitive' understanding of environmental and climate-friendly behaviour. Instead of black-boxing institutional and sociological explanatory factors, research attention should be directed to revealing the impact of institutional barriers and cultural norms. While doing so, the relevance of individual- and incentive-based factors should not be neglected. Indeed, our survey finds a weak connection between attitudes and actions. Our point, however, is to stress that other institutional and social factors play a more prominent explanatory role than economic incentives.

Instead of chasing individual behaviours, climate scholars and policy-makers need to shift their gaze and broaden their perspective. Research on climate action should investigate further the role of institutional capacity, social context and energy infrastructure in explaining action and failure of actions. Thus, we suggest that climate action studies replace the ABC behavioural model with one that takes account of the more multi-faceted understandings of the socio-political reality in which individual behaviour is embedded.

When climate change is addressed solely as a matter of individual agents, it neglects the fact that modern complex societies have a degree of institutional capacity that can be activated to deal with climate change. Institutional capacity is the ability of a society or institution to manage political challenges. Institutional capacity is a measure of the level of resilience of a society or institution.

If we want to slow climate change and counteract its harmful effects, it requires a break with the current dominant economic approach to our political concepts and ways of understanding political challenges. The capacity and resilience paradigm offers a framework in which individuals' actions are approached as mediated by a number of social and related factors (List & Pettit, 2011; Ostrom, 2012; Scavenius & Rayner, this volume). In other words, failures of action are not only something that should be investigated by asking people about their economic incentives and moral values. The cause of failures of action can also be found in the declining capacity of current democratic governance structures. Climate change is currently not adequately politically managed, which can be attributed to the economic policies that have resulted in extensive privatisation, deregulation and outsourcing. A consequence of this single-minded focus on economic policies is the erosion of political institutions' capacity and a general de-institutionalisation of key policy areas (Scavenius, 2014, Chapter 2, this volume; Economist Intelligence Unit, 2015). The formal political institutions still exist, but their capacity to decide, execute and implement comprehensive climate policies has been severely weakened. Empirical research on the causes and effects of the declining institutional capacity is needed. Survey studies are appropriate in order to understand individuals' self-reported ideas about incentives and values, but the survey studies are less suitable methods for comprehending the mediating explanatory factors of climate behaviour. Future research design should consider including methods other than survey studies.

Conclusion

The purpose of this chapter is to draw attention to the lack of correlation between people's green attitudes and people's green behaviour. The literature on environmental behaviour has discussed at length whether to explain green behaviour by referring to people's interior economic and psychological dispositions or people's exterior social, political and institutional surroundings. Our focus has centred on the value–action gaps theorised within the framework of a critical discussion of the causal and rational assumptions about behaviour.

Less attention has been paid by behavioural scholars to the action–value gaps that identify green acts conducted by people with few or no self-reported green values. In the Danish survey study, we find a significant but weak correlation between people's values and acts; most notably, we find evidence that people with low green values carry out almost the same amount of green acts as people with high green values.

On the basis of this study, we argue that the concrete political and institutional context, not people's green attitudes exclusively, determines their levels of green action. The explanation of why people conduct green actions cannot be found in traditional economic and rational models of behaviour. The double-gap finding indicates that the explanation of green actions should be found in non-attitudinal and non-personal factors. More research is obviously needed in order to understand the complex social and institutional dynamics that both impede and stimulate green action.

As long as climate research and climate policy-makers simply ask how we get individual actors to act in a more climate-friendly fashion, we maintain an action deficit. This is because the actors' freedom to act is subject to social and institutional barriers that currently make individual micro-actions irrelevant and ineffective in climate politics.

A paradigm shift would cope with the current action deficit and constitute a platform for developing climate subsidies. There is a need for climate research and climate policy to also engage in the facets of climate issues that are at an 'over-individual' level, where our political communities and societal institutions are interesting actors in their own right.

We suggest that further studies be conducted in order to develop a fully fleshed-out empirical understanding of the institutionalised mediating factors that explain climate action and failures of climate action. We opt for a new research design that breaks with the economic-rationalistic ABC model – one that experiments with methods that are not centred on answers provided by individual agents.

Note

1 $n = 1006$. Weighted by age and gender. The survey was conducted in December 2015.

References

Ajzen, I. & Fishbein, M. (1980). *Understanding attitudes and predicting social behaviour*. Englewood Cliffs: Prentice-Hall.

Ajzen, I. (1985). From Intentions to Actions: A Theory of Planned Behavior. In J. Kuhl & J. Beckman (Eds.), *Action-Control: From Cognition to Behavior* (pp. 11–39). Heidelberg, Germany: Springer.

Ajzen, I. (1991). The Theory of Planned Behavior. *Organizational Behavior and Human Decision Processes, 50*(1), 179–211.

Aldred, Jonathan (2016). Emissions Trading Schemes in a 'Non-Ideal' World. In Clare Heyward and Dominic Roser (Eds.), *Climate Justice in a Non-Ideal World* (pp. 148–168), Oxford: Oxford University Press.

Antonetti, P. & Maklan, Stan (2014). Feelings that Make a Difference: How Guilt and Pride Convince Consumers of the Effectiveness of Sustainable Consumption Choices. *Journal of Business Ethics, 124*(1), 117–134.

Bjurström, A., & Polk, M. (2011). Physical and economic bias in climate change research: a scientometric study of IPCC Third Assessment Report. *Climatic Change, 108*(1–2), 1–22.

Blake, J. (1999). Overcoming the 'Value-Action Gap' in Environmental Policy: Tensions Between National Policy and Local Experience. *Local Environment, 4*(3), 257–278.

Clapp, J., & Thistlethwaite, J. (2012). Private Voluntary Programs in Environmental Governance: Climate Change and the Financial Sector. In Karsten Ronit (Ed.), *Business and Climate Policy: The Potential and Pitfalls of Private Voluntary Programs.* New York: United Nations University Press,

Eurobarometer 300 (2008). *European attitudes towards climate change. Report.* Available at: www.ec.europa.eu/public_opinion/archives/ebs/ebs_300_full_en. pdf.

Flynn, R., Bellaby, P. & Ricci, M. (2009). The 'Value-Action Gap' in Public Attitudes Towards Sustainable Energy: The Case of Hydrogen Energy. *Sociological Review, 57*(2), 159–180.

Gausset, Q. & Hoff, J. (Eds.) (2015). *Community Governance and Citizen-Driven Initiatives in Climate Change Mitigation.* London: Routledge.

Gausset, Q., Hoff, J., & Scheele, C. (2015). Environmental Choices: Hypocrisy, Self-Contradictions and the Tyranny of Everyday Life. In Q. Gausset & J. Hoff (Eds.), *Community Governance and Citizen-Driven Initiatives in Climate Change Mitigation.* London: Routledge.

Guagnano, G., Stern, P., & Dietz, T. (1995). The New Ecological Paradigm in Social-Psychological Context. *Environment and Behavior, 27*(6), 723–743.

Gundelach, P., et al. (2012). *Klimaets sociale tilstand* [The social condition of the climate]. Aarhus: Aarhus Universitetsforlag.

Hines, J. M., Hungerford, H. R., & Tomera, A. N. (1987). Analysis and Synthesis of Research on Responsible Environmental Behavior: A Meta-Analysis. *Journal of Environmental Education, 18*(2), 1–8.

Hurley, S. (2011). The Public Ecology of Responsibility. In C. Knight and Z. Stemplowska (Eds.), *Responsibility and Distributive Justice.* Oxford: Oxford University Press.

Kollmuss, A., & Agyeman, J. (2001). Mind the Gap: Why Do People Act Environmentally and What Are the Barriers to Pro-environmental Behavior? *Environmental Education Research, 8*(3), 239–260.

List, C., & Pettit, P. (2011). *Group Agency: The Possibility, Design, and Status of Corporate Agents.* Oxford: Oxford University Press.

New York Times & Stanford University (2015). Resources for the Future. Poll on Global Warming. *New York Times,* 29 January 2015. Accessed 14 December 2016, www.nytimes.com/interactive/2015/01/29/us/politics/document-global-warming-poll. html

Ostrom, E. (2012). Nested externalities and polycentric institutions: must we wait for global solutions to climate change before taking actions at other scales? *Economic theory, 49*(2): 353–369.

Rayner, S. (1991). A Cultural Perspective on the Structure and Implementation of Global Environmental Agreements. *Evaluation Review, 15*(1), 75–102.

Sanne, C. (2002). Willing Consumers – or Locked-in? Policies for a Sustainable Consumption. *Ecological Economics, 42*(1–2), 273–288.

Scavenius, T. (2014). *Moral Responsibility for Climate Change. A Fact-Sensitive Political Theory.* Copenhagen: University of Copenhagen.

Scavenius, T. (2017). The Issue of No Moral Agency in Climate Action. *Journal of Agricultural and Environmental Ethics, 30,* 225–240.

Shove, E. (2010): Beyond the ABC: Climate Change Policy and Theories of Social Change. *Environment and Planning A, 42*(6), 1273–1285.

Stern, P. C. (2000). Towards a Coherent Theory of Environmentally Significant Behavior. *Journal of Social Issues, 56*(3), 407–24.

Thøgersen, J. (2005). How may consumer policy empower consumers for sustainable lifestyles. *Journal of Consumer Policy, 18,* 143–178.

Urry, J. (2008). Climate Change, Travel and Complex Futures. *The British Journal of Sociology, 59*(2), 261–279.

5 The social contract for climate risks

Private and public responses

W. Neil Adger, Tara Quinn, Irene Lorenzoni, and Conor Murphy

Introduction

Climate is experienced in individual places, and knowledge of it is constructed through narratives and prototypes of lived experience of weather, identity, culture and risk. The construction of risks is the result of the interaction between many individual perceptions, but also of institutions, habits and norms. In many senses, the institutional challenges of dealing with climate change are embedded in and constrained by beliefs and perceptions about the role of climate in society. Institutions are further constrained by the perceptions of who is responsible for action: the perceived role of state and citizens in dealing with climate change, allied to the role of collective action, autonomy and the agency of individuals.

In this chapter, we focus on the relationship between action taken by the state in the face of climate change and the role of the individual citizen. We do so in relation to the need to adapt to weather-related risks and a changing climate in particular. Action taken to address climate impacts across scales, sectors and actors results from particular and context-specific interactions: here we develop insights on the context-specificity of the institutional challenges of climate change. We lay out the importance of the interplay between government and citizen responses to climate change, and how the different actors involved shape the social contract around desirable planning objectives and response to events.

In effect, we argue that institutional capacity for dealing with climate change ultimately depends on the social licence for action by those affected by the government action. We show that in order to explain why governments act, or do not act, in the face of apparent risks from climate change, one needs to account for how much demand from the affected public there is for intervention. Second, where the government is judged by citizens not to have been sensitive to procedural legitimacy issues, citizens themselves are less likely to take adaptive measures in their own homes. Governments faced with climate change dilemmas need to be proactive in providing the space in which public demand and discussion for action can be articulated.

We use empirical examples from our own research on weather-related hazards to illustrate how different actors shape each other's behaviour, paying particular attention to the role of perceived fairness of state action in determining household-scale adaptation. We demonstrate that incorporating such insights leads to greater explanation of why climate politics can be ineffective.

Public and private governance of climate change

Managing climate change impacts and risks bears upon a multiplicity of institutions within and outside government, including formal and informal civil society, and private actors and markets. In varied combinations, they exert powerful influences on each other and the overall capacity to respond to climate change. Institutional characteristics and the relationships among them influence and affect responses to change against a backdrop of historical and cultural legacies. How decisions are made, who takes part in them, who is affected by them and how, inherently raise issues around social and environmental justice (Schlosberg, 2013).

In the negotiation of the social contract on climate change, there are trade-offs between objectives such as minimising vulnerability, maximising equity, and promoting system resilience. Eakin, Tompkins, Nelson, and Anderies (2009) showed how the prioritisation of key elements within climate strategies would inevitably require a shift in focus to some groups or communities within certain locations, leading to neglect of others; with deep and often overlooked temporal implications. The implementation of policies designated as climate change adaptation has already led to communities disaffected by decision-making protesting about lack of legitimacy. Throughout the world, communities have resisted the implementation of infrastructure projects, the building of dams, afforestation projects and the resettlement of populations that have been justified and sold as climate change adaptation (McDowell, 2013; Kothari, 2014).

Increased consultation and direct citizen participation in decision-making are often advocated as means to build opportunities for the legitimacy of decision-making processes. Coastal and flood planning, disaster planning and urban design processes experiment with new forms of deliberation and citizen science. Widening participation in decision-making on controversial or high-risk issues is promoted as a way to improve decision-making due to the wider knowledge base; to enhance inclusivity, give greater legitimacy to the outcomes, and reduce vulnerability to risk by increasing citizens' competence in understanding and addressing those risks (e.g. Pearce, 2005). Such efforts are widespread across the world. Yet the expectations of increased citizen participation have in many senses failed to materialise. Powerful institutions direct decisions to their own desired outcome, and consultation processes bow to those shouting the loudest. Realistically, participatory processes for planning for climate change most often take place in situations where power relations may be so strong as to preclude any fair and open debate (Few, Brown, & Tompkins, 2007).

Continued societal involvement may become even more urgent in the future, as risks are amplified and negative synergies between them emerge. Risks to food security, to public health and to displacement potentially act together and create unforeseen outcomes and increased levels of vulnerability (Dilling, Daly, Travis, Wilhelmi, & Klein, 2015).

Yet individuals and communities engage in debate about future risks that are pertinent and salient (Corner, Markowitz, & Pidgeon, 2014). When we analysed how diverse members of the UK lay public talk about climate change, we found that they gravitate towards moral arguments. In focus group discussions held across the UK in 2013, members of the lay public in places ranging from Glasgow to Brighton were asked about their support for interventions to help people vulnerable to the impacts of climate change. Analysis of the focus group discussions demonstrated, for example: widespread discussion of solidarity with victims; need and entitlement to help with adaptation; and duties to protect from harm (Adger, Butler, & Walker-Springett, 2017). In effect, climate change adaptation is frequently framed in moral terms, and when it is discussed in that light, fairness becomes salient.

Social contracts involve expectation of other parties. Expectations of what governments do for their citizens and expectations of institutional intervention directly affect how individuals respond to change. This expectation is always clear around the time of crisis. Evidence on the effect of expectations of government on subsequent behaviour is widespread: an example is the study of whether people returned to New Orleans or relocated elsewhere after the impacts of Hurricane Katrina in 2005 (Chamlee-Wright & Storr, 2010). That study clearly shows that it is expectation of performance of government and capacity and scope of government that determined how and why different populations acted in the aftermath of that traumatic event.

Climate change impacts will be place-based; climate change will be experienced tangibly in places that matter to local populations. Personal affective closeness to a location or its features helps define the sense of place and identity of an individual or community. Place attachment affects individuals' preferences for or aversion to future change. In real lives and real places where weather-related disasters strike, dealing with insurance can be a significant negative stress and impact on well-being (Carroll, Morbey, Balogh, & Araoz, 2009; Whittle, Walker, Medd, & Mort, 2012). The impact of negotiating with insurance companies is one of the major traumas following a flood event, for example, but when coverage is rescinded this directly affects how people assess their own vulnerability in terms of where they live. Positive place attachment is tied up with continuity and rootedness. Insurance is an important component of this feeling, especially in recovery following disruptive events. This shifting relationship of responsibility often translates at the individual level into a significant change in affective relationship with place. Hence, place, identity and the relative role of citizen and public governance of climate change are intimately intertwined.

The case of flood risk

Climate risks are real and multiplying. All risks, from direct impacts on health to complex issues of food security, have private and public elements, involve the private sector, central and transnational governance and impinge on place, identity and expectations of government through their social contract with citizens. We illustrate the governance dilemmas by focusing on flood risk. River and coastal floods are among the most costly natural hazards (UNISDR, 2011). In 2015 floods affected more than 12.5 million people and accounted for three of the top ten disasters in terms of number of deaths (EM-DAT, 2016). While at regional scales trends in the magnitude and timing of floods are difficult to detect from observations, increases in major flood events have been detected and are expected to continue (Milly, Wetherald, Dunne, & Delworth, 2002). The science of event attribution asserts that there is already a discernible human fingerprint in the risk of widespread fluvial flooding (Pall, Aina, Stone, Stott, Nozawa, Hilberts, et al., 2011; van Oldenborgh, Otto, Haustein, & Cullen, 2015).

While floods are an element of climate governance, even without climate change, flood risk will increase over the coming decades as populations expand and exposure increases. Increases in global temperature are expected to cause an intensification of the global water cycle, with consequent increases in flood risk. In parts of the world projected to have wetter winters, soil moisture will likely increase; with greater catchment wetness there will be an increasing likelihood of flood events. Where annual floods are driven by snowmelt, warmer temperatures will likely alter the timing of floods and increase their magnitude, while sea level rise increases riverine flood risk in coastal areas. Arnell and Gosling (2014) estimate that by mid-century, the range in increased exposure to river flooding at higher-level climate scenarios is 31–450 million people and 59,000 to 430,000 km^2 of cropland. Winsemius, Aerts, van Beek, Bierkens, Bouwman, et al. (2016) project absolute damage from floods globally to increase by up to a factor of 20 by the end of the century without adaptive action. The scale of such impacts will challenge governance systems everywhere.

The relation between flooding and climate is inevitably complex: disaster burden is not influenced by weather and climate events alone, with population growth, wealth, and changes in vulnerability also being major drivers (Visser, Petersen, & Ligtvoet, 2014). Either way flood risk is only expected to increase over coming decades and will be a key element of climate change adaptation planning across scales. Institutional arrangements determine the extent to which adaptive capacities for flooding may be mobilised in the civic, private and public sector. Many national capacities in flood management require radical institutional reforms to overcome complacency and lack of engagement with land use, development and other drivers of risk (Eakin et al., 2011, Wilby & Keenan, 2012).

Expectations and adaptive capacity

We have studied the implications of expectations, social contracts for adaptation and public demand for change through observations of how they evolve in real time. The study of how political systems interact with those at the sharp end of risk is best done by examining lived experiences of events and examining their aftermath in the present day. Observational studies provide direct evidence and potentially robust insights into processes that will be important as the risks themselves change with the climate.

In November 2009, a series of Atlantic depressions led to the formation of an atmospheric river that caused heavy rainfall events across the UK and Ireland: in particular in Cumbria, North West England and in Galway, western Ireland. The same meteorological event and its subsequent social impacts were, in effect, a natural experiment: how do impacts unfold against the social context of these two different places? What are the significant elements in shaping responses? The study focused on the perceptions of those directly impacted by the floods about public authority response, what they felt about future flood risk, and what they themselves were willing to do to manage flood risk. Our results showed that the social contexts in the two places resulted in the flood events unfolding along different trajectories and that underpinning this divergence was the difference in the social contract between citizen and states in Ireland and in England.

This study on flood risks surveyed 358 households across four towns both in Cumbria and in Galway. In each county the surveys captured the experience both of people directly affected by floods, and those who lived in the flood-affected towns but did not experience flooding in their households. To get under the skin of the processes of risk and the perceptions of the capacity of the political system to cope, we used mixed quantitative and qualitative methods: the details can be found in Adger et al. (2013, 2016).

The study confirmed, as one would expect, significant differences between perceptions of individual responsibility in the two jurisdictions, and also a divergence in expected future risk by the two directly affected populations. People affected in both Ireland and England had similar expectations that governments are responsible for reducing the likelihood of flooding in the future. Whilst there was agreement that householders also share some responsibility for future flood risk management, a significantly larger element of the Cumbrian respondents were willing take on personal responsibility relative to those in Galway. This divergence in willingness to engage in protective action was also reflected in residents' willingness to adapt their homes to flood risk.

Do expectations of government affect how individual citizens act? Flooded Cumbrian respondents were significantly more likely to be willing to make physical changes to their homes than those who had been flooded in Galway. At the same time those flooded in Galway were significantly more likely than their non-directly affected community members to believe that they would experience flooding again in the next five years.

What do these results show about the adaptive capacity of government? We interpret these results in the light of the social contract between citizens and states: differences between the two contexts were shown in this instance to be related to citizen experience of fair treatment by their government. Other dimensions of the social contract include expectations of protection from future flood risk by public bodies and personal willingness to take preventative measures at the household scale. In Galway poor government performance in terms of procedural and distributive actions resulted in poor trust in the ability of government to prevent flooding in the coming years.

So in two different countries, the same event had very different outcomes in terms of the social contract for protection from harm. Table 5.1 highlights how the social contracts diverge in Ireland and England, both for how the demand for policy change is articulated, and in how the social contract leads to a level of individual citizen responsibility. In effect, the lessons from the Cumbria floods in England fed directly into public advocacy for renewed planning and investment in flood risk. The UK has continued to live with floods, and flood events of winters 2013/2014 and 2015/2016 became national political crises. These have focused on public expectations of solving the causes of flooding, compensation for those affected, and attempts to achieve long-term solutions in flood provision. In Ireland, the legacy of the 2009 floods was more litigation and blame of the government, coupled with a disillusionment about its capacity to enact effective national planning.

The observed lack of trust in public bodies revealed in our detailed survey work was coupled with a sense of helplessness that was reflected in Galway by residents' lack of willingness to engage in personal action to prevent flooding. In contrast, whilst the social contract in Cumbria was far from perfect, previous experience of floods and agency response led to higher levels of trust in government to protect citizens and also a higher level of belief that public agencies had

Table 5.1 How the demand for policy change and individual behaviour is affected by flood events: UK and Ireland compared

	UK	*Ireland*
Demand for policy change	UK government continues to devolve responsibility for flood risk. Populations at risk of flooding demand clarity on proposed new configurations of insurance and government partnership on management of flood risk.	Contested liability: Litigation on blame. Individuals took Irish government to central European Court of Justice. Questioning of relationship between government and developers – distrust in distribution of power and calls for separation of interests.
Changing expectations of government responsibility and individual responsibility	Higher willingness to adapt and recognition of individual responsibility and long-term risk. Expectations remain that government would continue to provide protection.	Increased disillusionment with government agencies and lowered expectations of efficient and fair intervention. Helplessness in face of flood risk, lack of willingness to personally undertake adaptive measures.

Source: Derived from Adger et al. (2016).

acted fairly. This faith in government was matched by personal engagement in adaptive measures.

Responses to questions on fairness pointed to some of the underlying mechanisms that were informing the different responses and expectations of householders in the two jurisdictions. In Galway, less than half of respondents agreed that everyone in their community had received help promptly following the floods, whereas three quarters of Cumbrians felt that this had been the case. The same number of Cumbrians agreed that resources had been distributed to those who needed them the most, but in Galway over half of respondents felt that this had happened in the post-flood period.

In Figure 5.1 we summarise the mechanisms that affected how willing the UK and Ireland populations are to act to take individual responsibility for adaptation. The figure shows the principal significant factors that affect sampled respondents' willingness to act: perceived fairness of process, fairness of outcome and prior knowledge of risks explain whether individuals will act themselves or expect governments to act on their behalf in the face of risk. The results show that how governments perform makes a significant difference to citizen action. Perceptions of fair process (who was consulted), perceptions of fair outcome (how assistance was distributed), and the perceived capacity of government, as well as prior experience, all affect individual willingness to act. In Ireland, however, expectations of government were lower, and perceived fair process was not a significant factor, compared to whether flood victims perceived that agencies 'did all they could'.

These results confirm that expectations of government play a significant role in citizen response to natural disasters (also see Chamlee-Wright & Storr, 2010).

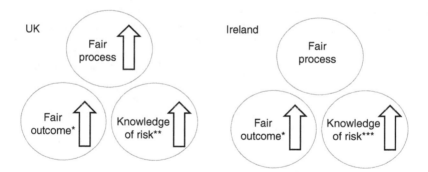

Figure 5.1 Willingness to take individual action on climate risks is affected by perceptions of fair process, fair outcome and by prior experience and knowledge of flood risk.

Source: derived from Adger et al. (2016).

Notes
Based on sampled respondents experiencing flooding in Ireland and Galway.
Arrows indicate statistically significant factors.
*   'Authorities did all they could'.
**   'Prior awareness of risk'.
*** 'Perceived government capacity'.

Previous experience of the functioning of a social contract and projected expectations shape the individual sense of duty and associated behaviours. Context is key; social contracts unfold in response to the specific jurisdictional circumstances in which they apply.

Local adaptation

The example we elaborated above of the individual citizen and public elements of climate governance illustrates a number of themes. First, private willingness to take action to respond to climate risks, in this case flooding, is related to perceived procedural and distributional fairness, and the political context in which they operate. The capacity to adapt to changing circumstances results from previous experience of the event (current as well as historical) as well as the perception of externally driven responses to the event. Second, the role of institutions, their visibility and perceived fairness of actions is key in enabling individuals to feel supported in their contribution to enhancing personal and collective capacity to adapt. There is increasing documentation of conflict in adaptation planning and resistance to the implementation of plans, not least in governments that seek to move and resettle whole populations, justified as climate change adaptation (de Sherbinin, Castro, Gemenne, Cernea, Adamo, Fearnside, et al., 2011). Populations directly affected will have much to say about planning into the future.

Renegotiation of the social contract

The international climate regime illustrates how competences and contracts matter. The anticipated impacts are highly diverse across space. The Paris climate agreement enshrines the rediscovered worth of nationally defined and led contributions to mitigation, opening spaces and creating opportunities for more closely dovetailing with adaptation efforts, traditionally inspired and crafted based on local circumstances and needs. Although the Paris Agreement is, in effect, a framework, there have been positive commitments from individual countries. The Paris Agreement has changed expectations of what international-led action is capable of facilitating, enhancing and maintaining in response to climate change. But international diplomacy is not the sole or even the most important competence for effective governance of climate change. Core legitimacy and effectiveness at the national level matter. Creating national climate policy, however, is challenging, as demonstrated in our study of flood risk.

The principal challenges arise because national climate policy deals with climate change impacts generated through actions taken both within and outside jurisdictions, as well as natural climate variability.

The idea of the scope and competence of national governance systems is often articulated as a set of relations between government systems and the authority and legitimacy they maintain. This set of relations is analysed through social contract theory. Such theory seeks to explain how governments and responsibility evolve over time as emerging risks pose challenges to the established consensus concerning the role of the state. Clearly the balance of power between civil society and the state is not fixed. And the nature of the state has changed irrevocably over time, not least in relation to the ability of states to deal with issues that cross national boundaries – ranging from the threat of nuclear Armageddon, to global pandemics, to the flow of capital and the rise of transnational corporations.

Governments essentially need to protect vulnerable populations, whether from economic circumstances and poverty, or from public health threats. So, too, do they need to protect populations from climate change risks, even if those risks are not known or widely perceived to be critical. Political scientists have argued that environmental risks create completely new roles for states and expand the role of government, and they call for governments to consult more widely, adopt precautions, and engage citizens in new ways of democratic deliberation (Dryzek, et al., 2002).

So what does civil society expect from governments in terms of protection from climate change? The social contract between citizens and states is limited: it can exclude those who may not recognise the legitimacy of governments; it can emerge from less-than-legitimate lobbying among key actors; and it cannot directly represent citizens of the future (Weale, 2011). Climate risks are particularly problematic for consensus building for government because of uncertainty and uneven distribution of burdens (Pelling, 2011). And if climate change is experienced as a set of new risks through weather extremes (changes in floods, wildfire, drought and heatwaves), then it is in the politics of dealing with catastrophe that the social contract and sets of expectations are reformed.

Hence, we suggest that extreme events can have significant roles in both small regulatory changes and in large political upheavals (Pelling & Dill, 2010; O'Brien et al., 2011; Adger et al., 2013). Renegotiations of social contracts are likely to be a primary mechanism for both adaptation and transformation. Our work on flood risk suggests that the method and substance of consultation matter, and that in particular procedural legitimacy is important in ensuring that individual citizens and the state enter into synergistic relationships for climate change adaptation.

Response capacity

Over the past two decades the scientific community and the policy world has slipped into a comfortable distinction between climate change mitigation and adaptation: mitigation addresses decarbonisation of the economy while adaptation deals with the consequences of climate change risks on society as they begin to bite.

This dichotomy has facilitated some standardised assumptions about the scale of the challenge and the distribution of responsibility. Mitigation is often framed as a global problem requiring a consensus among nation states to provide a global public good, while adaptation is regarded as a national or even local problem, often involving the protection of private interests. However, this sharp dichotomy between mitigation and adaptation seems less relevant in the light of the shift to multi-level and polycentric approaches to mitigation as represented in the Paris Agreement. While it would be futile to rely solely on individual citizen action to reduce greenhouse gas emissions on the scale required to curtail global warming, the ability to implement any kind of greenhouse reduction measures depends on the institutional capacity of public authorities and jurisdictions to mobilise and coordinate public action. Both mitigation and adaptation require positive behavioural change that will simply not emerge unless the public has the capability to act, trust in government motivations, and lend support to interventions that they perceive to be fair in their implementation (Lorenzoni, Nicholson-Cole, & Whitmarsh, 2007). Public support and public protest over fracking throughout Europe and North America attest to this finding. Hence we would argue that climate governance involves common principles for transformation and that attempting to analyse adaptation and mitigation as separate policy domains misses how cause and effect in climate change are intertwined in most mental models of the issue (see also Termeer et al., 2017). In effect the potential scale, scope and interconnectedness of many climate change risks will require radical transformation in economic and social structures (Kates, Travis, & Wilbanks, 2012).

From the perspective of mobilising behavioural change and support for climate policies, the dichotomy between mitigation and adaptation is unhelpful. Long time-frames, scientific uncertainty about impacts and about social and economic futures all conspire to constrain climate policy in the arenas of both mitigation and adaptation to risks. While they may be separated in terms of political responsibility, the challenges of building public trust and legitimacy are common to both. Response capacity according to Tompkins and Adger (2005) is the set of options in which governance deals with climate change dilemmas and in which trade-offs between mitigation and adaptation are implicit. Hence the societal, geographical and political circumstances in which institutions operate are the key to a new politics of climate change. These circumstances determine the willingness and ability of individual actors and of collective action to make sense of and act on climate change. The ability and willingness of different parties is a matter of negotiation and expectation – a social compact or social contract between governments and the governed.

Many climate risks, particularly those that impact on the natural world, involve significant public goods: nature conservation and public infrastructure all require collective action and investment to realise their adaptation benefits for all. Hence while the public and individual remits for mitigation of greenhouse gas emissions and for adaptation to the impacts of climate change are likely to have radically different mixes of public and private benefits, the governance of both arenas shares many of the same challenges and dilemmas. Interventions in either domain inevitably involve redistribution of benefits and risks, with support from those gaining, and resistance from perceived losers.

Discussion

This chapter focuses on the institutional context of dealing with the climate as it changes – the adaptation challenge. The governance of adaptation is polycentric in nature and happens at scales that do not necessarily fit its associated ecological and physical contours of risks. Adaptation has emerged in its own right as a policy field (Massey & Huitema, 2016). The challenges of adaptation include the multi-jurisdictional nature of climate change impacts and therefore of multi-level governance. Further, there are trade-offs between infrastructure and policy solutions with their long-term sustainability, including trade-offs and synergies between so-called mitigation and adaptation. For adaptation, there is a diversity of policy instruments, ranging from private insurance to public regulation and investments in infrastructure. Governments are faced with the standard dilemmas of the optimal mode of governance that both distributes risk fairly, and is seen to be legitimate for the risks involved, many of which are intangible or in the future.

Much effort in assessing the impacts from climate change has focused on the distribution of risks across space and across society. Many studies have demonstrated who is vulnerable to risks and why. In general, socially marginalised groups, whether marginalised by income, class, gender, or by their location in geographic peripheries, are more vulnerable to the impacts of climate change. Vulnerability is heightened among elderly populations globally who are physiologically at risk from extreme heat, while populations on low incomes, especially in countries with insufficient public infrastructure, are vulnerable to food insecurity, displacement risks and from the impacts of weather-related hazards in urban settings. Hence the governance of adaptation to climate change inevitably focuses on the social processes of reducing such vulnerability and minimising impacts. It is also clear that adaptation will not happen smoothly and costlessly, but rather in relation to existing economic and political structures that determine who and what is presently vulnerable and whose values count (Adger & Barnett, 2009).

Conclusion

We have outlined both theoretical arguments and our experience in generating evidence on the public and private benefits and risks associated with climate change. The public and individual citizen mix is constructed by rules, expectations and politics, with significant consequences for climate justice, the distribution of vulnerability, and ultimately for the incentive to take climate change seriously as a global issue.

All the evidence, for both climate adaptation and mitigation, demonstrates that the effectiveness of climate action is dependent on the synergy between collective state action on climate risks and the expectations of citizens and individuals about the collective risk and their individual responsibility. There is also emerging evidence that perceptions of fairness and the performance of governments in times of crisis further influence the effectiveness of climate action.

The institutional capacity for dealing with climate change involves collective competences and individual action. Most public discourse on climate change includes discussions on what 'governments should do' to promote individual behaviour that is low-carbon and sustainable and/or takes account of climate risks. Ultimately, however, governments themselves are constrained in their ability to steer action within their jurisdictions by the culture and politics of the governed.

Acknowledgements

The research was supported by the Tyndall Centre for Climate Change Research, the UK Economic and Social Research Council (Grant ES/M00687/1) and the Irish Environmental Protection Agency (TRANS-ADAPT Grant 2014-CCRP-MS.15). We thank the editors for helpful direction and participants at the Challenges of Climate Politics conference at the University of Copenhagen in June 2013 for stimulating discussion resulting in this chapter.

References

Adger, W. N., & Barnett, J. (2009). Four reasons for concern about adaptation to climate change. *Environment and Planning A, 41*, 2800 2805.

Adger, W. N., Butler, C., & Walker-Springett, Kate (2017). Moral reasoning and adaptation to climate change. *Environmental Politics, 26*, 371–390.

Adger, W. N., Quinn, T., Lorenzoni, I., & Murphy, C. (2016). Sharing the pain: perceptions of fairness affect private and public response to hazards. *Annals of the Association of American Geographers, 106*, 1079–1096.

Adger, W. N., Quinn, T., Lorenzoni, I., Murphy, C., & Sweeney, J. (2013). Changing social contracts in climate change adaptation. *Nature Climate Change, 3*, 330–333.

Arnell, N. W., & Gosling, S. N. (2016). The impacts of climate change on river flood risk at the global scale. *Climatic Change, 134*, 387–401.

Botzen, W. J. W. (2013). *Managing Extreme Climate Change Risks through Insurance.* Cambridge: Cambridge University Press.

Carroll, B., Morbey, H., Balogh, R., & Araoz, G. (2009). Flooded homes, broken bonds, the meaning of home, psychological processes and their impact on psychological health in a disaster. *Health and Place, 15*, 540–547.

Chamlee-Wright, E., & Storr, V. H. (2010). Expectations of government's response to disaster. *Public Choice, 144*, 253–74.

Corner, A., Markowitz, E., & Pidgeon, N. (2014). Public engagement with climate change: the role of human values. *Wiley Interdisciplinary Reviews: Climate Change, 5*, 411–422.

de Sherbinin, A., Castro, M., Gemenne, F., Cernea, M. M., Adamo, S., Fearnside, P. M., Krieger, G., Lahmani, S., Oliver-Smith, A., Pankhurst, A., & Scudder, T. (2011). Preparing for resettlement associated with climate change. *Science, 334*, 456–457.

Dilling L., Daly, M. E., Travis, W. R., Wilhelmi, O. V., & Klein, R. A. (2015). The dynamics of vulnerability: why adapting to climate variability will not always prepare us for climate change. *Wiley Interdisciplinary Reviews: Climate Change, 6*, 413–425.

Dryzek, J. S., Hunold, C., Schlosberg, D., Downes, D., & Hernes, H. K. (2002). Environmental transformation of the state: the USA, Norway, Germany and the UK. *Political Studies, 50*, 659–682.

Eakin, H., Eriksen, S., Eikeland, P. O., & Øyen, C. (2011). Public sector reform and governance for adaptation: implications of new public management for adaptive capacity in Mexico and Norway. *Environmental Management, 47*, 338–351.

Eakin, H., Tompkins, E. L., Nelson, D. R., & Anderies, J. M. (2009). Hidden costs and disparate uncertainties: trade-offs in approaches to climate policy. In Adger, W. N., Lorenzoni, I., & O'Brien, K. (Eds.), *Adapting to Climate Change: Thresholds, Values, Governance* (pp. 212–226). Cambridge: Cambridge University Press.

EM-DAT (2016). The OFDA/CRED – International Disaster Database. Université Catholique de Louvain, Brussels, Belgium. Available at www.emdat.be

Few, R., Brown, K., & Tompkins, E. (2007). Climate Change and coastal management decisions: Insights from Christchurch Bay, UK. *Coastal Management, 35*, 255–270.

Johnson, C. L., Tunstall, S. M., & Penning-Rowsell, E. C. (2005). Floods as catalysts for policy change: Historical lessons from England and Wales. *Water Resources Development, 21*, 561–575.

Kates, R. W., Travis, W. R., & Wilbanks, T. J. (2012). Transformational adaptation when incremental adaptations to climate change are insufficient. *Proceedings of the National Academy of Sciences, 109*, 7156–7161.

Kothari, U. (2014). Political discourses of climate change and migration: resettlement policies in the Maldives. *Geographical Journal, 180*, 130–140.

Lorenzoni, I., Nicholson-Cole, S., & Whitmarsh, L. (2007). Barriers perceived to engaging with climate change among the UK public and their policy implications. *Global Environmental Change, 17*, 445–459.

Massey, E., & Huitema, D. (2016). The emergence of climate change adaptation as a new field of public policy in Europe. *Regional Environmental Change, 16*, 553–564.

McDowell, C. (2013). Climate-change adaptation and mitigation: Implications for land acquisition and population relocation. *Development Policy Review, 31*, 677–695.

Milly, P. C. D., Wetherald, R. T., Dunne, K. A., & Delworth, T. L. (2002). Increasing risk of great floods in a changing climate. *Nature, 415*, 514–517.

O'Brien, K. (2012). Global environmental change II: from adaptation to deliberate transformation. *Progress in Human Geography, 36*, 667–676.

O'Hare, P., White, I., & Connelly, A. (2016). Insurance as maladaptation: resilience and the business as usual paradox. *Environment and Planning C: Government and Policy, 34*, 1175–93.

Pall P., Aina, T., Stone, D. A., Stott, P. A., Nozawa, T., Hilberts, A. G., Lohmann, D., Allen M. R. (2011). Anthropogenic greenhouse gas contribution to flood risk in England and Wales in autumn 2000. *Nature, 470*, 382–386.

Park, S. E., Marshall, N. A., Jakku, E., Dowd, A. M., Howden, S. M., Mendham, E., & Fleming, A. (2012). Informing adaptation responses to climate change through theories of transformation. *Global Environmental Change, 22*, 115–126.

Pearce, L. (2005). The value of public participation during a hazard, impact, risk and vulnerability analysis. *Mitigation and Adaptation Strategies for Global Change, 10*, 411–441.

Pelling, M. (2011). *Adaptation to Climate Change: from Resilience to Transformation.* London: Routledge.

Pelling, M., & Dill, K. (2010). Disaster politics: tipping points for change in the adaptation of sociopolitical regimes. *Progress in Human Geography, 34*, 21–37.

Schlosberg, D. (2013). Theorising environmental justice: the expanding sphere of a discourse. *Environmental Politics, 22,* 37–55.

Termeer, C. Dewulf, A., & Biesbroek, G. R. (2017). Transformational change: governance interventions for climate change adaptation from a continuous change perspective. *Journal of Environmental Planning and Management, 60,* 558–576.

Tompkins, E. L., & Adger, W. N. (2005). Defining response capacity to enhance climate change policy. *Environmental Science and Policy, 8,* 562–571.

UNISDR (2011). *Global Assessment Report on Disaster Risk Reduction.* Geneva: United Nations International Strategy for Disaster Reduction.

van Oldenborgh, G. J., Otto, F. E., Haustein, K., & Cullen, H. (2015). Climate change increases the probability of heavy rains like those of storm Desmond in the UK – an event attribution study in near-real time. *Hydrology and Earth System Sciences Discussions, 12,* 13197–13216.

Visser, H., Petersen, A. C., & Ligtvoet, W. (2014). On the relation between weather-related disaster impacts, vulnerability and climate change. *Climatic Change, 125,* 461–477.

Weale, A. (2011). New modes of governance, political accountability and public reason. *Government and Opposition* 46, 58–80.

Whittle, R., Walker, M., Medd, W., & Mort, M. (2012). Flood of emotions: emotional work and long-term disaster recovery. *Emotion, Space and Society, 5,* 60–69.

Wilby, R. L., & Keenan, R. (2012). Adapting to flood risk under climate change. *Progress in Physical Geography, 36,* 348–378.

Winsemius, H. C., Aerts, Jeroen C. J. H., van Beek, Ludovicus P. H., Bierkens Marc F. P., Bouwman, Arno et al. (2016). Global drivers of future river flood risk. *Nature Climate Change, 6,* 381–385.

6 The role of civil society actors in climate change adaptation

Jens Hoff

Introduction

Hurricane Sandy hit New York City around 7.30 pm on 29 October 2012. A number of factors contributed to making it the worst natural disaster ever to hit New York. Apart from the unusual angle with which the hurricane hit the coast, the arrival of the storm coincided with high tide in New York Harbor. The effect of the storm tide was devastating. Along the shoreline, it smashed buildings and engulfed entire communities. It flooded roads, subway stations and electrical facilities, paralysing transportation networks and causing power outages that plunged hundreds of thousands into darkness. It left 43 New Yorkers dead, 6,500 patients were evacuated from hospitals and nursing homes, 90,000 buildings were in the inundation zone, 1.1 million New York children were unable to attend school for a week, close to two million were without power, and there was an estimated US$19 billion of damage (PlaNYC, 2013, p. 11).

Even though Sandy has been called a 'game-changer' (IV 2:3[1]), and has certainly changed New York forever, the hurricane did not find the city wholly unprepared. Since 2007, when the City of New York launched its first long-term sustainability plan: *A Greener, Greater New York*, the first in the PlaNYC-series, the city has made regular sustainability plans including climate change adaptation measures. It had a Climate Change Adaptation Task Force and it was just about to give PlaNYC an overhaul when hurricane Sandy hit (IV 2:2). The need for such an overhaul became painfully acute as Sandy clearly demonstrated the inadequate character of the earlier plans.[2]

Below, we discuss in detail the climate change adaptation plans that came out after Sandy. As we will see, these plans are quite precise concerning the initiatives taken and the agencies responsible for the action. However, this is true only to a lesser extent when it comes to the role that civil society actors are imagined to play in the formulation as well as the implementation of the initiatives. This is even more surprising as the academic literature is increasingly stressing both the need to focus at the local level when dealing with climate change adaptation, as well as on the engagement of civil society actors in order to optimise response capacity[3] in any given community (Wamsler & Brink, 2014; Adger et al., 2004; Lutz, 2008).

In New York City attention to this point is also aptly demonstrated by experience. Thus, in the aftermath of Sandy some of the affected neighbourhoods recovered much quicker than others. Comparing, for example, Manhattan's Lower East Side with the nearby Financial District shows that the latter recovered much faster. This is most likely a result of the higher income levels as well as the economic importance of the Financial District (IV 2). Thus, the combination of good personal financing insurance and support from the City, eager to get the key economic centre of the city up and running, made a quick recovery possible. Additionally, the greater affluence of the local individuals and businesses allowed them to more easily utilise their capital to assist recovery and implementation of new resiliency effort to reduce vulnerability in the case of future disaster events (IV 10).

So, for theoretical as well as empirical reasons, it is interesting to look at the resiliency plans and practices in New York City in a civil society perspective. I deliberately use the term 'civil society perspective' to stress that this perspective is different from a citizen and a community perspective. Thus, we read the academic literature as stressing the necessity of engaging *all* civil society actors in building response capacity; i.e. citizens, community groups and NGOs as well as businesses, think tanks, academic institutions, etc.

Plans are strategies for future action and, as they progress, it is possible to look at the process through which they progress, and eventually their outcomes. As Pelling (2011) points out, strategies can be seen as intentions, actions (or process) and outcomes. Much of the outcome of the PlaNYC plans remains to be seen, so we will limit our analysis to the two first strategic dimensions: intentions and actions.

New York City consists of very diverse communities and boroughs with widely different compositions of residents in terms of ethnicity, employment status, educational level and household income. For the City administration to communicate and collaborate with all these groups and interests, therefore, constitutes a tremendous challenge. Analysing how this challenge has been handled makes New York City a particularly interesting and information-rich case, warranting the label 'extreme case' (Flyvbjerg, 2001). Choosing an extreme case puts our theories to the test, and applying them in this context hopefully pushes their limits, thereby contributing to theoretical development.

The role of civil society actors

Most existing theories on climate change adaptation and resilience building deal with the role of civil society actors in one form or another (Pelling, 2003, 2011; Pelling et al., 2015; Smit & Wandel 2006; Huq & Burton, 2003;Wamsler, 2015; Adger et al., 2009; Tyler & Moench 2012), and all see citizen or civil society participation as a necessary element in adaptation strategies and resilience building. However, with a few exceptions (e.g. Wamsler & Brink, 2014) these theories are not very precise when they specify the exact interplay between civil society actors and relevant public authorities, and when they characterise different types of civil society actors. For this reason, I find it necessary to build a more detailed framework for understanding participation of civil society actors in climate change adaptation, which I call the 'interactionism approach'.

On the basis of their analysis of a number of cases of climate change mitigation, Hoff and Gausset (2016, p. 3) have developed a two-dimensional model, where the vertical axis presents a continuum showing that adaptation initiatives can be driven either primarily top-down from an authority or be civil society driven, bottom-up. The horizontal axis presents a continuum showing that initiatives can target either individuals, social groups (households, housing cooperatives, etc.) or communities. Nikpour and Lendal (2016, pp. 40–41) have improved the model by breaking out the subcategories of the vertical continuum, inspired by Arnstein's (1969) original ladder of participation. Authority-driven initiatives are (law) enforcement initiatives, awareness initiatives (one-way communication) or consultation initiatives (welcoming citizens' views, but final decisions rest with authorities). Civil society initiatives can be partnerships (civil society actors have direct involvement in decision-making processes and implementation), delegated responsibility (resources and responsibility is delegated to civil society), or civil society driven (stakeholders have the ideas, carry out the projects and only rely on authorities for advice and support) (see Figure 6.1).

An important conclusion in Hoff and Gausset's (2016) examination of a number of case studies is that initiatives driven top-down by authorities and targeting individual behaviour (square 1) are not as effective as initiatives driven

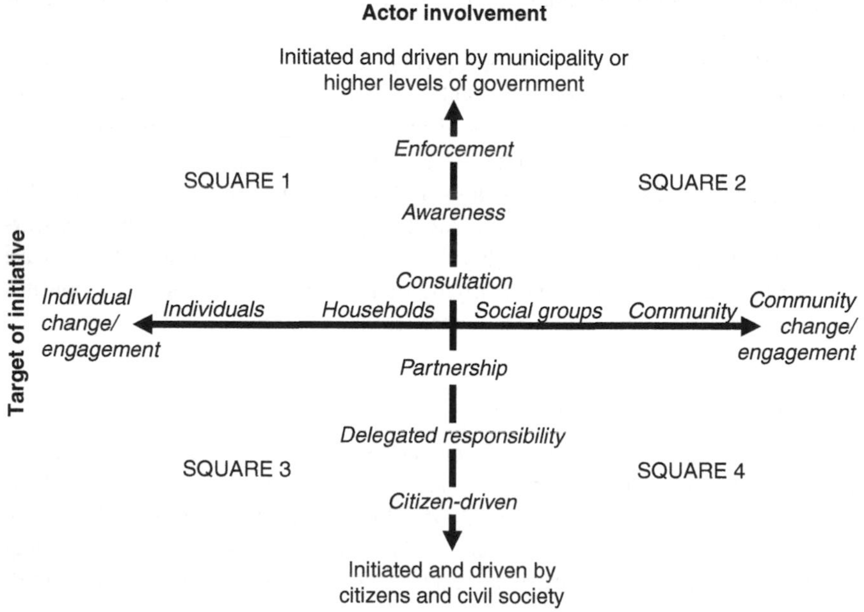

Figure 6.1 Civil society initiatives can be partnerships, delegated responsibility, or civil society driven.

by groups of citizens or local communities (square 4) in addressing climate change challenges. The latter initiatives seem to be able to generate a deeper and broader level of change as new behavioural standards emerge and dominant social norms start to change. This approach, as well as that of Nikpour and Lendal, resonates well with other analyses of the interface between citizens and authorities in climate change adaptation. Wamsler and Brink (2014, pp. 69, 75) for example stress that authorities can choose to support citizens by either (1) directly reducing the risk they face, or (2) strengthening the ability to reduce the risk themselves. The latter demands (extended) interactions between citizens and authorities. Wamsler and Brink analyse a number of such interactions in Sweden and find that citizens' adaptive practices can compensate for the lack of assistance from authorities, can hinder it, or can show the extent of reliance on it.

Equally, assistance from authorities can support, discourage or obstruct citizens' adaptive practices. This points towards a possibly conflictual relationship between civil society actors and authorities, leading Brink and Wamsler (2016, p. 24) to propose a revision of the Hoff and Gausset model, replacing the horizontal axis with a continuum running from 'contested actions' to 'collaborative actions'.

Nikpour and Lendal (2016, pp. 42ff.) have also elaborated on the authority–civil society interactions, and combine facilitation by authorities with engagement by civil society in what they call a 'resilience cycle'. Nikpour and Lendal argue that facilitation and engagement can take place at a micro level, such as between a municipal department and a concrete initiative from civil society, and at a macro level, where the municipality in the broadest terms provides access and incentives for civil society stakeholders to engage. The resilience cycle shows that initiatives towards resilience might come from local government, but could also come from civil society itself. If the initiative comes from a local government, there are typically a number of barriers in the implementation process, which are related to either the capacity of agents, the structure and workings of institutions or the structure of the political/economic system.

Nikpour and Lendal use the resilience cycle to construct a typology of barriers to resilience building. The aim is to draw attention to the fact that a local government and civil society agents are faced with identical types of barriers. Using Tyler and Moench's (2012) distinction between agents, institutions and core city systems, Nikpour and Lendal point towards barriers concerning the possession of economic and political assets, such as money, resources, housing, political network, etc. *Institutional* barriers can be of many kinds; for example, limited decision-making power or competencies of different authorities, lack of relevant information or lack of trust in authorities (Yang, 2005). Barriers to a city's *core systems* can be many, and may either be adjacent to it or remote. Nikpour and Lendal extend the notion of 'system' to also include what they call 'root causes', which, following Pelling (2011), they see as wider and less visible factors that 'lie in social, cultural, economic and political spheres, often overlapping and interacting' (Pelling, 2011, p. 86). The root causes will often have to do with the structure of the dominant regime. In our analysis below, we will be aware of different types of barriers that might impede or hinder the active engagement of civil society actors in resilience-building activities in New York City.

Political intentions

Looking at the types of collaboration between civil society actors and the political-administrative system, it makes sense to distinguish between intentions; i.e. the types of collaboration described in the different resiliency plans, and actions; i.e. the collaborations described in the section above on decision-making and implementation. In the following, I focus first on the differences between two political plans, and thereafter on the implementation side of the policy.

The first sustainability plan to be published after Sandy was 'A Stronger, More Resilient New York' (June 2013), which focused solely on resiliency measures. A progress report came out in 2014, but the next substantial report was 'One New York. The Plan for a Strong and Just City' published in April 2015 (New York, 2015). A progress report on this plan was published in 2016. I focus on the two main plans, as they were published under the direction of two different mayors and, therefore, might represent two different approaches to resilience building. In addition, the second report might reflect that a learning process on resilience has been taking place.

'A Stronger, More Resilient New York' was compiled by the Special Initiative for Rebuilding and Resilience (SIRR), convened by Mayor Bloomberg just after Sandy. The aim of the Special Initiative was to analyse the impacts of the storm on the city's buildings, infrastructure and people, to assess the risks the city faces from climate change in the medium term (2020s) and long term (2050s), and to outline ambitious yet achievable strategies for increasing resiliency citywide. In addition, the Special Initiative was asked to develop proposals for the areas hardest hit by Sandy to help them to improve safety (PlaNYC, 2013, pp. 1–2). For this reason, the report is divided in two parts. After an initial description of climate analysis in New York City and the initiatives taken to improve the quality of climate analysis, the first part deals with citywide infrastructure. It deals meticulously with areas such as coastal protection, buildings, insurance, utilities, healthcare, transportation, etc. and support systems such as food supply and solid waste. The second part of the report deals with the areas most affected by Sandy.

After describing the area, including its socio-economic characteristics, the report recounts what happened during Sandy, makes a risk assessment, and outlines the initiatives to be taken to increase the resilience of the area. However, it deviates from the chapters in part one as it describes how local communities are engaged, and how the conversations with these groups (formal and informal) helped the Special Initiative to clarify the priorities for the area. An example is from the Brooklyn-Queens Waterfront where the Special Initiative met with and briefed around 20 elected city, state and federal officials on a monthly basis, and with the five Community Boards in the area and the 40+ faith-based, business and community organisations every 4–6 weeks.

It is clear that the main thrust of the plan is to ensure continuation of existing systems' functions into the future. This has necessitated some alterations in institutions and organisations, such as establishing the Special Initiative and the Mayor's Office of Recovery and Resiliency, but these novelties do not seem to impinge much on the work of existing departments. The adaptation strategy represented by this plan is, therefore, what is normally known as a 'mainstreaming' strategy (Smit & Wandel, 2006), meaning that resiliency measures are built into, or on top of, existing plans and strategies.

'One New York, The Plan for a Strong and Just City' (New York City, 2015), which was Mayor Bill de Blasio's first sustainability plan, is structured very differently from 'A Stronger, More Resilient New York'. It is centred on four principles and four visions for the city. The principles are said to have formed the goals and initiatives of OneNYC, and are quite similar to the core challenges and opportunities in the report (New York City, 2015, p. 26). These are a growing population, infrastructure needs, an evolving economy, urban environmental conditions and climate change, and growing inequality. The four visions are centred on growth, equity, sustainability and resilience, and they are further subdivided into a number of areas for which concrete goals are set, and initiatives developed. A concrete example is the resilience area for which goals are set for neighbourhoods, buildings, infrastructure and coastal defence.

OneNYC can also be said to represent a mainstreaming strategy. This is stressed by the report insofar as it is made clear that it builds on the foundation of a number of strategies, launched during the first year of the de Blasio administration, in such areas as schools, housing, greenhouse gas reductions, job creation, and poverty (OneNYC, 2015, pp. 12–13). In addition, no new units are set up or announced to take care of the initiatives at the City level, meaning that this is a job for existing departments or units. A significant effort was made to hear citizens concerning their priorities and ideas via online and telephone surveys, community meetings and meetings with civic organisations (New York City, 2015, pp. 18–20). In contrast to 'A Stronger, More Resilient New York', the report also touches on the need to engage community-based organisations on a more permanent basis. An example is the resiliency initiative to strengthen community-based organisations, which starts by recognising that social infrastructure plays an important role in making communities ready for the unexpected. Therefore, the city will work to build capacity in communities by strengthening community-based organisations that serve their neighbours, and by working to expand civic engagement and volunteerism. This is done by, among other things, strengthening community-based organisations' services, information capacity, and ability to conduct community-level emergency and resiliency planning (New York City, 2015, p. 224).

Through its focus on building capacity in local communities, and its special focus on rising inequality, the plan promises to lift 800,000 New Yorkers out of poverty or near poverty by 2025. Thus, the plan can be said to represent a more holistic approach than 'A Stronger, More Resilient New York'. Attention is paid not only to infrastructure and environment, but also to such areas as economy, health and well-being, and the interdependencies between the areas when it comes to building resilience. The plan therefore represents what Pelling (2011) calls a 'transitional approach' (as opposed to a resistance or a transformation approach).

Following the description above, it is clear that there is a marked difference between the two resiliency plans written only two years apart. Where Bloomberg's plan has a narrow focus on increasing the resistance of the city to climate hazards, especially flooding, in the short to medium term, de Blasio's plan has a more holistic approach with a broader perspective on resilience and with a longer time perspective. While it seems clear that the shift reflects a shift in political priorities, it is less clear whether it also reflects a learning process by city authorities. We will dig deeper into this question in the analysis below, where we also move from *intentions* to the *implementation* processes.

Differences in economic resources

An important aspect of building resilience in communities and societies is the access of citizens, community groups, local businesses, etc. to key resources and the capacity to use these resources skilfully when exposed to hazards. Tyler and Moench (2012), for example, stress that, for civil society actors, the capacity to mobilise assets, the capacity to organise and re-organise, and the capacity to learn from earlier failures is crucial for resilience building. However, the possibilities for utilising these capacities are dependent on the institutions that frame them. The two most important institutions in this respect are the market, determining prices for food, housing, education, insurance, etc. and the political-administrative apparatus able to distribute rights and resources to different groups, and to sanction or reward different types of behaviour authoritatively.

In the introduction, I gave an example of how the affluence of local individuals and businesses in Manhattan's Financial District allowed them to utilise their capital to assist recovery and implementation of new resiliency measures to reduce vulnerability in the case of future disaster events more easily than individuals and businesses in the nearby Lower East Side area (IV 10). This confirms the trend found by Smit and Wandel (2006, p. 287) that high vulnerability is correlated with low income and high (ethnic) diversity.

Complementary to, or derived from, capital assets and income are other key resources such as housing, education and health services. With a growing population, affordable housing is a huge issue and the same goes for good and affordable education, as well as health services. According to a survey conducted in relation to OneNYC, housing and education are the two most important issues to be addressed by City government (20% and 29% of interviewees mention these as most important) while health services come in third (7%). One main reason for lack of access to these resources is the great inequality in incomes in New York City. Inequality is higher than the national average (OneNYC, 205, pp. 30–31) and has been growing since the recession in 2008. Of all New Yorkers, 21.5% live below the poverty line, with another 23.6% close to that line. In addition, homelessness is at a record high. This lack of, or fragile access to, key resources for almost half of the population in New York City is a factor that severely limits resilience building in the city.

Differences in political resources

Additional to economic assets are political assets; i.e. on the one hand, access to political decision-making bodies and, on the other hand, the capacity to use this access to further one's preferences. New Yorkers have different possibilities for access to political decision-making bodies. First, they can elect the Mayor and the City Council every four years, as well as a Public Advocate, a Comptroller, and Borough Presidents. The 51 members of the City Council are elected in 51 council districts in the five boroughs of New York (Brooklyn, Queens, Manhattan, Bronx and Staten Island) via a 'first-past-the-post' system. The Borough Presidents preside over a Borough Board, which, besides the President, consists of the City Council members from the borough and the chair of each of the Community Boards. Community Boards are appointed by the Borough President and can consist of up to 50 members. There are altogether 59 Community Boards in New York City. These boards have only advisory power, but can advise on the important issue of land use and zoning; an issue which is known to galvanise the New York electorate.[4] The office of Borough President is said to be mostly a ceremonial role, but Borough Presidents effectively control land use and zoning in the Boroughs due to their advisory role in ULURP (Uniform Land Use Review Procedure) and their work with the resourceful NYC Economic Development Corporation.[5]

This formal political system creates quite a few possibilities for access to political decision-making for New Yorkers. Many New Yorkers are apparently ready to use the system, as there are 4.5 million registered voters in NYC (Board of Elections website, 2015). However, at the latest election in 2013, turnout was only 26%. This is the lowest point in a long downturn in turnout at local elections in New York City. Looking back at the 1950s and 1960s average turnouts were around 70–80%. While there are probably several explanations for this drop, what we see is that New Yorkers are increasingly not using the formal political system to promote their preferences, and there seems to be a disconnect between a 'political elite' and the population at large; a phenomenon known throughout the Western world (Bang & Dyrberg, 2003, pp. 232–234). We do not have data to substantiate a hypothesis that persons of colour, persons not in the labour force, with low education and low income are overrepresented among the non-voters, but this is most likely if the voting pattern in New York City follows the patterns known from other contexts.

This situation diminishes the political capital of New Yorkers, particularly the least privileged, and must be seen as detrimental to resilience building in the City. One of the attempts to remedy this situation is the strengthening of community-based organisations proposed in OneNYC. Whether this will increase access to political decision-making, particularly for less privileged groups, remains to be seen.

Apart from the formal political system, access to political-administrative decision-making and implementation can also take place via informal channels such as contacts and meetings with key persons in the political and administrative system. The interviews informing this chapter give an impression of how these channels are used, but of course not nearly a complete overview. The interviews suggest that these channels are used by all sorts of advocacy groups, NGOs and businesses (IV 2, IV 8). An example is the New York City Environmental Justice Alliance (NYCEJA): a citywide non-profit operation that operates as a network connecting grassroots groups that work with environmental and climate change matters in an equity perspective. There are seven groups altogether working in deprived areas of the city. The network was created in 1991, and is run by a board, but its mission and priorities are set by the groups working on the ground. They describe themselves as doing advocacy work, policy analysis and research (IV 9:2). They depict their contacts with the political-administrative system in the following way:

> We work at different levels. So there [are] like, conversations we will have with the Mayor or staff, so depending on the issue it could be Deputy Mayor or Economic Development and Housing, it could be the environmentally related staff or directly with the agencies in terms of specific projects, specific collaborations, data sharing for some of our work, or providing comments as parts of technical advisory committees. And then the other more important constituent in that conversation is the City Council, who is often requesting comments from us to inform their positions on city actions and city responsibilities.
>
> (IV 9:10)

Apart from this collaboration with the political-administrative system on a day-to-day basis, there are also two concrete examples of the political influence of NYCEJA. One is from just after Sandy hit, where the network co-convened a group called the Sandy Region Assembly, which developed an agenda called the Sandy Regional Assembly Recovery Agenda. The document was given to the Special Initiative for Rebuilding and Resilience, and some of its recommendations were incorporated in Bloomberg's 'A Stronger, More Resilient New York City' report (PlaNYC, 2013). The other example is Bill de Blasio's OneNYC report, which connects the issues of sustainability and equity in addressing resilience: something NYCEJA had been advocating for a number of years (IV 9:3–4). Another organisation successfully using the informal channel is the Metropolitan Waterfront Alliance, a nine-year-old NGO focusing on the NYC harbour.[6]

One might speculate whether these advocacy groups/NGOs exaggerate the extent of their contacts with and influence on city administration in order to appear more important than they really are. However, their role in resilience building/city planning is largely confirmed by the administration.

Business, academia, think tanks and private foundations are also among the civil society/non-state actors that are regularly consulted by the administration or who demand to be part of the consultative circuit. Concerning businesses, their contact is institutionalised in different ways. One way the administration (Dept. of City Planning) is collaborating with business is through Civic Advisor Communities, which include business representatives. The City administration is here working with representatives of a neighbourhood's commercial owners and, on occasions, some actual business owners directly. Another way the administration is collaborating with business is through the NYC Economic Development Corporation. However, working with businesses might be at least as difficult for city administration as working with community and advocacy groups; and the process is prone to contain elements of conflict. In coordinating resiliency building, making public and private entities work together constitutes a challenge, among other things because elements of the infrastructure are privately owned.

Private foundations and academia are also a part of the consultation circuit. In New York City the Rockefeller Foundation has played a large role both through its development of the 100RC methodology (a methodology for building resilience spread to 100 cities around the world) and network of which New York City is a member, and through co-funding with the federal Department of Housing and Urban Development (HUD) the Rebuild by Design competition. Part of this competition was won by the Danish architect firm, BIG, which has developed a project called 'The Big U': a city development project meant to enhance resiliency around the southern tip of Manhattan.[7] As part of developing the project, BIG spent quite some time meeting with the local citizens, who are very diverse as the 10 miles being developed contains some of the most underserved communities at Manhattan; the lower East Side and Chinatown as well as the more affluent areas such as the Financial District and the West Side. Therefore, businesses here are doing their own citizen consultations. When BIG was asked whether this entails dealing with conflicting interests, the answer was:

> No, you listen to everyone and then you prioritise. You list everything; some things go against what somebody else said, but people I think are extremely intelligent in these situations and scenarios. You will demonstrate through workshops, listening and reports just how you are digesting the information, how you are prioritising, and then the choices that have to be made with the budget.
>
> (IV 7:6)

A representative from NYC Environmental Justice Alliance was less convinced about the lack of conflict around 'The Big U':

> So the community was not happy. The first place (of development) is not going to benefit the community that is most vulnerable, that had been

participating in all the meetings…. It will actually stop a couple of blocks away from the Two Bridges Area, which has the highest concentration of low-income and population of colour on the lower East Side.

(IV 9:12)

This situation epitomises a fundamental conflict in resiliency building in New York City, which is between real estate developers and communities. When older, industrial areas along the waterfront are re-zoned, new residential and commercial buildings appear, thus spurring gentrification. One example is Williamsburg, which lost 20% of its Latino population between 2000 and 2010 (40% in some districts) (IV 9:11). The city administration also acknowledges that there are conflicts between commercial and community interests that relegate these conflicts to the political arena.

Academia is another civil society actor that is playing a prominent role in the consultation circuit. In the case of the New York City Panel on Climate Change (NPCC), this role has even been formalised. The first panel (NPCC1), consisting of academics and private sector experts with knowledge on climate change, infrastructure, social science and risk management, was established in 2008. Its role was made formal through Local Law 42, which established the city panel as an ongoing body serving the City of New York by giving advice to the Mayor's Office of Sustainability and (later) the Mayor's Office of Recovery and Resiliency on climate change mitigation and adaptation. After Sandy the panel was reestablished in 2013 (NPCC2). Its latest report, focusing on resiliency, is from 2015 (Rosenzweig & Solecki, 2015). This input from the academic world is listened to and seen as very important by political-administrative planners and decision-makers.

We have seen from the examples above that a plethora of civil society/non-state actors try to exert their influence through informal channels on the political-administrative system or are sought-after partners of the system due to their financial resources, special knowledge, or ability to legitimise political-administrative projects or interventions. Another way to express this is that the informal channels have become the dominant arena of politics; politics is no longer primarily about input legitimacy through elections but output legitimacy that concerns the implementation side of the political system. The upside of this situation is that the policies being implemented are more effective and legitimate than if all these actors had not been a part of what I have called the 'consultation circuit'. The downside of the situation is that the influence exercised by the civil society actors in this arena does not compensate for the lack of influence through formal political channels, as it does not ensure equal, democratic representation for all. Thus, it is still a situation where the least resourceful risk losing out.

Citizenship involvement

'A Stronger, More Resilient New York' describes how local communities have been engaged, and the conversations that the Special Initiative for Rebuilding and Resilience has had with these communities when setting priorities in the plan. However, there is no mention of more long-term engagement of local communities or other civil society stakeholders, or of any strategies to strengthen capacities at this level. The type of actor involvement in the plan therefore does not move beyond what is called (raising) awareness and consultation in the citizenship ladder figure (see Figure 6.1).

'One New York, The Plan for a Strong and Just City' goes somewhat further in its attempts to involve civil society actors. First, a stronger effort was made to consult citizens about their needs and priorities, and second, as part of building resiliency the report suggests strengthening community-based organisations by enhancing their services, information capacity, and ability to conduct resiliency planning. This moves the type of actor involvement towards what is called 'partnership' and 'delegated responsibility' in Figure 6.1. However, it is not clear in the plan to what extent the city is actually aiming to fund these activities, and to what extent it will actually delegate more rights to community-based organisations[8] (following the responsibilities) in the interests of resilience building. For this reason, most of the actor involvement described in 'A Stronger, More Resilient New York' falls into Square 2 of Figure 1, while that of 'One New York' falls into Square 4 near the horizontal axis.

Looking at actor involvement in the political-administrative system through informal channels, we find that many advocacy groups, NGOs as well as business, academia and private foundations have close relations to the system, and take part in discussions, consultations, etc. on a regular basis. However, this relationship seldom moves beyond the 'consultation' phase. Only in a few cases has the relationship been institutionalised, moving the collaboration towards what can be characterised as a 'partnership'. For example, the New York City Economic Development Corporation was established to link industry and the city while the City Panel on Climate Change integrates academia into the mitigation and resiliency efforts of the city.

We conclude that neither of the New York City resiliency plans reaches the higher rungs on the 'actor involvement ladder'. Even though a lot of attention is given to local communities, including their businesses, etc. this has not resulted in a real delegation of more responsibilities to these communities or in more and better facilitation of citizen-driven initiatives.

Summing up, we can conclude that at the institutional level the low levels of support for community-based groups and little facilitation of citizen-driven initiatives are major barriers to building resilience. The Community Boards seem to have too few competencies and resources to be able to act as an important force in (local) resilience building. Furthermore, there might be a problem in the 'double mandate' of the City Council members. Thus, the City Council members are also members of the relevant Borough Council, which tends to make them very much representatives of the borough when acting in the City Council. According to one observer (IV 2), it makes New York City appear as 'a collection of villages' rather than a unified city. This makes concerted action to build resilience more difficult.

At the agency level, i.e. citizenship level, there is little doubt that the biggest barrier to resilience building is the inequity in the access to different kind of resources, both material and political, among individuals and communities. Above, we gave examples of how this inequity resulted in different levels of vulnerability in different neighbourhoods.

At the system level, we already mentioned the lack of delegation of responsibilities and facilitation of citizen-driven initiatives as a barrier. Other barriers at this level are the difficulties in coordinating the actions of public and private owners of critical infrastructure such as energy, transport, shelter, waste and telecommunication. Finally, one can point to a (local) political economy, which creates huge differences in incomes and wealth, leaving around 45% of the population in New York below or near the poverty line.

Conclusion

The aim of this chapter has been twofold: first, to construct a theoretical framework that can be used to analyse the role of civil society actors in climate change adaptation strategies, and second, to demonstrate the usefulness of this framework through an analysis of the resiliency plans and practices in New York City.

We called the constructed theoretical framework the 'interactionism approach'. The contribution of this approach was first of all to give us a nuanced tool to analyse actor involvement and, second, it also delivered a typology for barriers to resilience building using the distinction between systems (or root causes), social agents and institutions. Departing from three different sources of data the chapter went on to analyse the role of civil society actors in resilience building in the city of New York.

Using the theoretical framework to characterise the involvement of civil society actors in the city's resiliency plans, we found that the plans do not move decisively beyond the 'consultation' type of actor involvement. Looking more specifically at actions/processes, we saw that even though a lot of attention is given to local communities, including their businesses, etc. this has not resulted in a real delegation of more responsibilities to these communities or in more and better facilitation of citizen-driven initiatives. We argued that neither in terms of intentions (i.e. the different resiliency plans of the city) or in action (informal actor involvement) does the building of resiliency in the city reach the higher rungs on the 'actor involvement ladder'. This is a barrier to resilience building in New York City at the system level.

The conclusion is thus clear: even though Sandy was a disaster and it has been a window of opportunity for a renewal of urban planning and the establishment of new relations between communities and the political-administrative system, it remains to be seen to what extent the opportunities will be used to benefit the majority of New Yorkers. The purpose of this chapter has been to draw attention to the gap between political intentions of citizenship involvement and resilience building on the one hand and the political realities in New York

City. Attention has been drawn to the fact that economic and political inequality is a barrier to resilience building and thus to effective climate change adaptation strategies.

Notes

1 The data used in this chapter come from three sources: the first source is the various sustainability and resiliency reports released by the City of New York. The four that will be used here are: 'A Stronger, More Resilient New York' (PlaNYC New York City, 2013); 'Progress Report 2014. A Greener, Greater New York, A Stronger, More Resilient New York' (PlaNYC New York City, 2014); 'One New York. The Plan for a Strong and Just City' (New York City, 2015) and 'One New York. 2016 Progress Report' (New York City, 2016). The second source are 11 semi-structured interviews with key employees in City and Borough administration, representatives of (local) NGOs and community groups, business representatives and representatives from think tanks and academia. The interviews were conducted between 12 March and 8 May 2015 by the author, and typically lasted 35 to 60 minutes. In the text, the interview persons arc rcfcrrcd to with a number and a reference to the page in the interview. Thus, a reference might be IV 2:2, meaning that the reference comes from page 2 in the interview with interview person number 2. The third source of data is reports, articles, newspaper clippings, policy briefs, etc. partly given to the author by the NGO's and community groups interviewed and partly found on the Internet. This is supplemented with material on other community groups in NYC found on the Internet. An example is the Red Hook Initiative, working especially to empower youth, which played an important role locally in the aftermath of Sandy.
2 There were many problems with the earlier plans, some of them to do with data coming from federal sources. A pertinent example is that the flood plain maps for New York had not been updated since 1983. These maps, coming from FEMA (the Federal Emergency Management Agency) indicated that in the event of a so-called '100-year flood' (a flood estimated to have only a 1% chance of happening in any given year) 33 square miles of New York City might be inundated. However, Sandy flooded 51 square miles or 17% of the city's total land mass, exceeding the 100-year floodplain boundaries by 53%. Much of the city's critical infrastructure was within the flooded areas – including hospitals and nursing homes, key power facilities, transportation networks and all the city's wastewater treatment plants.
3 In defining 'response capacity' I am inspired by Tompkins and Adger (2005), who, departing from the two critical elements of the availability and penetration of new technology, and the ability or willingness of society to change, propose that *response capacity is ... the ability to manage both the causes of environmental change and the consequences of that change* (Tompkins & Adger, 2005, p. 564). I take that to mean that response capacity is the sum of the adaptive and mitigative capacity of a given community (Yohe, 2001), and that a community can be seen as resilient when its response capacity is sufficient to fend off all or most of the social and natural hazards it might be exposed to in a foreseeable future.
4 See for example *New York Post*, 9 July 2013.
5 NYC Economic Development Corporation is a powerful organisation, which has the responsibility to manage NYC's real estate. It is legally a private company, but wholly owned by the city, and with a CEO appointed by the Mayor (IV 10:1).
6 Its mission is to make the harbour cleaner, the water quality better, make the harbour more accessible for people, maintain jobs and build resilience. It has managed to create a constituency for the waterfront by bringing together big maritime interests, environmentalists, recreational boaters, engineers, landscape architects, etc. and works as an umbrella organisation for around 835 different groups.

7 The project will build 10 miles of coastal defence that stretches from West 57th St to East 40th St (IV 7:1).
8 As we saw above, the formal Community Boards already have some rights in terms of, for example, their right to be heard concerning land use and zoning. Also, the Department of City Planning has staff assigned to the Community Boards that appear at Community Board meetings every month. So there is a close relationship between the Department and each of the Community Boards *'intentionally so because we want to have eyes and ears on the ground as to what is happening, and what are the concerns of the communities, and be able to respond when issues get raised'* (IV 8:9). Whether these boards are going to play a more important role or not in resilience building in the future is not discussed in any of the plans.

References

Adger, N. W., et al. (2004). *New Indicators of Vulnerability and Adaptive Capacity.* Norwich: Tyndall Centre for Climate Change Research.

Adger, N. W., et al. (2009). Adaptation Now. In Adger, N. W., I. Lorenzoni, & K. O'Brien (Eds.), *Adapting to Climate Change: Thresholds, Values, Governance* (pp. 1–22). Cambridge: Cambridge University Press.

Arnstein, S. R. (1969). A ladder of citizen participation. *Journal of the American Institute of Planners, 34*(4), 216–233.

Bang, H. P., & Dyrberg, T. B. (2003). Governing at close range: demo-elites and lay people. In Bang, H. P. (Ed.), *Governance as social and political communication* (pp. 222–240). Manchester: Manchester University Press.

Board of Elections (2015).

Flyvbjerg, B. (2001). *Making Social Science Matter. Why social inquiry fails and how it can succeed again.* Cambridge: Cambridge University Press.

Hoff, J. & Gausset, Q. (Eds.) (2015). *Community Governance and Citizen-Driven Initiatives in Climate Change Mitigation.* London & New York: Routledge, Earthscan.

Huq, S. & Burton, I. (2003). Funding Adaptation to Climate Change: What, Who and How to Fund. *Sustainable Development Opinion.* London: IIED.

Lutz, W. (2008). *Forecasting societies' adaptive capacities to climate change (FutureSoc) Annex 1 – description of work.* Funded Research Proposal, European Research Council, Advanced Investigators Grant. Unpublished document. Laxenburg, Austria: International Institute for Applied Systems Analysis (IIASA).

PlaNYC New York City: 'A Stronger, More Resilient New York'. PlaNYC report, June 2013.

PlaNYC New York City: 'Progress Report 2014. A Greener, Greater New York, A Stronger, More Resilient New York'. PlaNYC Report 2014.

New York City: 'One New York. The Plan for a Strong and Just City'. NYC, April 2015.

New York City: 'One New York. 2016 Progress Report'. NYC, 2016.

New York Post (2013): Why are boro presidents important? Jobs. http://nypost.com/2013/09/07/why-are-boro-presidents-important-jobs/ . Accessed 3 August 2017.

Nikpour, A., & Lendal, A. S. (2016). *Resilience-Building in a Resource Poor Kingdom. The challenges of municipal facilitation and resident engagement in Amman.* Unpublished MA. Thesis. Dept. of Political Science, University of Copenhagen.

Pelling, M. (2003). *The Vulnerability of Cities. Natural disasters and social resilience.* London & New York: Routledge, Earthscan.

Pelling, M. (2011). *Adaptation to Climate Change. From resilience to transformation.* London & New York: Routledge.

Pelling, M., et al. (2015). Adaptation and transformation. *Climatic Change, 133*(1), 113–127.

Rosenzweig, C. & Solecki, W. (2015). Volume 1336: Building the Knowledge Base for Climate Resiliency: New York City Panel on Climate Change 2015 Report. *Annals of the New York Academy of Sciences.*

Smit, B., & Wandel, J. (2006). Adaptation, adaptive capacity and vulnerability. *Global Environmental Change, 16*, 282–292.

Tompkins, E. L., & Adger, N. W. (2005). Defining response capacity to enhance climate change policy. *Environmental Science and Policy, 8*, 562–571.

Tyler, S. & Moench, M. (2012). A framework for urban resilience. *Climate and Development, 4*, 311–326.

Wamsler, C. (2015). Mainstreaming ecosystem-based adaptation: transformation towards sustainability in urban governance and planning. *Ecology and Society, 20*(2), 30.

Wamsler, C., & Brink, E. (2014). Interfacing citizens' and institutions' practices and responsibilities for climate change adaptation. *Urban Climate, 7*, 64–91.

Yang, K. (2005). Public Administrators' Trust in Citizens: A Missing Link in Citizen Involvement Efforts. *Public Administration Review, 65*(3), 273–285

Yohe, G. (2001). Mitigative capacity – the mirror image of adaptive capacity on the emissions side. *Climate Change, 49*, 247–262.

Part III
Institutional capacity in society

7 Institutional challenges of climate geoengineering[1]

Steve Rayner and Peter Healey

Introduction

Institutions are hard to define – so difficult in fact that the very idea has been described as an *essentially contested concept* (Gallie, 1955) that can only be clarified through argumentation. Broadly, across the social sciences, the term is applied to 'organizations with leaders, memberships, clients, resources and knowledge and also to socialized ways of looking at the world as shaped by communication, information transfer, and patterns of status and association' (O'Riordan, Cooper, Jordan, Rayner, Richards, Runci, et al., 1998, p. 346). Writing in the Blackwell Encyclopedia of Political Science, Gould (1991, p. 290) defines an institution as a 'locus of regularized or crystallized principle of conduct or, action, or behaviour that governs a crucial area of social life that endures over time'. For some theorists (e.g. Giddens, 1986; Smith, 1988) the idea of temporal stability looms large in their definitions, but for others even durability is not a defining factor. For example, Thompson (2008, p. 51) describes an institution as 'any non-randomness in behaviour (or indeed in the beliefs and values that are used to justify that behaviour)'.

Specifically in the context of climate change, O'Riordan et al. argue that

> Climate policy is shaped by formal organisational structures, as well as by informal networks of communication that, in turn, are products of values, norms and expectations. These institutions range from the formal deliberating bodies engaged in treaty making to the informal liaisons among policy analysts and policy executives, regulatory agencies, and the day-to-day activities of billions of people.
>
> (O'Riordan et al., 1998, p. 346)

The formation and delivery of climate policy thus involves institutional, and inter-institutional, values and capacities, interacting at a global scale.

Below we draw on the results of a major interdisciplinary research effort – Climate Geoengineering Governance (CGG) – that we designed and led to examine how existing institutional values and capacities play out in relation to a nascent issue in climate policy: the governance of climate geoengineering.

We also consider the developments that may be needed in institutional processes and capacities for the technical research that would have to be conducted before the potential contribution of particular geoengineering approaches to climate policy can be fully assessed.

Geoengineering

Defined by Britain's Royal Society as 'the deliberate large-scale manipulation of the planet's environment to counteract climate change' (Shepherd, Caldeira, Cox, Haigh, Keith, Launder, et al., 2009, p. 1), climate geoengineering is receiving growing attention from scientists, scholars and policy-makers concerned that the Paris Agreement to limit global average temperature rise to 1.5–2° Celsius cannot be achieved through conventional emissions mitigation alone (Anderson, 2015; Rau & Greene, 2015; Tollefson, 2015; UNEP, 2016).

The Royal Society Report – one of the most influential analyses in framing subsequent academic and policy discourse – distinguishes two principal mechanisms for moderating the climate by geoengineering. One is by reflecting some of the sun's energy back into space to reduce the warming effect of increasing levels of greenhouse gases in the atmosphere. This is described as solar radiation management or SRM. The other approach is to find ways to remove some of the carbon dioxide from the atmosphere and sequester it in the ground or in the oceans. This is called carbon dioxide removal or CDR. While SRM is seen as potentially fast-acting and having high leverage on global average temperatures, it is also widely regarded as controversial, requiring some kind of formal global agreement before implementation (Robock, 2012; Bodle & Oberthuer, 2014; Hulme, 2014a), which some consider impossible to achieve because of the probable uneven spatial distribution of SRM impacts on climate (Ricke, Morgan, & Allen, 2010). On the other hand, while it seems that land-based CDR could be adequately regulated by the domestic laws of individual nation states, it is likely to take a long time to ramp up and to require massive investment in infrastructure for which there is little incentive in the absence of a significant carbon price. Both SRM and CDR present significant challenges of institutional design.

The Royal Society report also recognised, but gave less prominence to, another way of discriminating between geoengineering technologies, that cuts across the distinction between SRM and CDR. Both SRM and CDR can be achieved by two different means.

One is to put something into the air or water or on the land's surface to stimulate or enhance natural processes. For example, injecting sulphate aerosols into the upper atmosphere imitates the action of volcanoes, which we know to be quite effective at reducing the sun's energy reaching the earth's surface; hence this is one candidate SRM technique. Similarly, we know that lack of iron constrains plankton growth in some parts of the ocean. It has therefore been suggested that adding iron to these waters would enhance plankton growth, taking up atmospheric CO_2 in the process. This would be a potential CDR technique also achieved by imitating nature.

The other approach to both SRM and CDR is through more traditional hard or 'black-box' engineering. Mirrors (either large or more likely myriad small ones) in space, either in orbit or at the so-called Lagrange point between the earth and the sun, would be a way of reflecting sunlight (SRM), while a potential CDR technique would be to build machines to remove CO_2 from ambient air and inject it into old oil and gas wells and saline aquifers in the same way that is currently proposed for carbon capture and storage (CCS) technology. Based on previous experience with emerging technologies, such as GM crops, one might expect that releasing novel substances into the atmosphere or oceans may prove more controversial than the engineering of discrete industrial-type facilities. On the other hand, perceived parallels between captured CO_2 and nuclear waste in Germany[2] suggest that the removal and storage of atmospheric carbon by machines might also attract opposition.

Nevertheless, combining these two means (natural systems enhancement and black-box engineering) with the two goals described above (SRM and CDR) creates a serviceable typology for discussing the range of options being considered under the general rubric of geoengineering (Table 7.1).

As Table 7.1 suggests, geoengineering proposals encompass a wide range of heterogeneous technical practices relying on diverse mechanisms, and operating in diverse media. Furthermore, with few exceptions there are no fully developed, tested and demonstrated geoengineering technologies on offer that can be seen to make a potential contribution to climate policy alongside mitigation and adaptation; and none whose efficacy and potential side-effects are fully understood. So, while the Royal Society's broad definition of 'deliberate large-scale manipulation of the planet's environment' is widely accepted, the question of whether or not any particular technology counts as geoengineering – rather than mitigation or adaptation – is often subject to dispute.

Equally contested is the appropriate level and means of governance and regulation for any particular technology at any stage in its development and possible deployment. Even experimental work on many of them (but not all) is controversial and, crucially, likely to involve activities and consequences that cross institutional and even international jurisdictions. Deployment of any of them would have to be considered and implemented in relation to climate change mitigation and adaptation efforts, as well as problems such as food shortages, lack of disaster resilience, migration, etc. in which anthropogenic climate change will be only one contributory cause.

Table 7.1 Geoengineering for climate change

	Carbon dioxide removal	Solar radiation management
Ecosystems enhancement	e.g. Ocean iron fertilisation, enhanced weathering	e.g. Stratospheric aerosols, marine cloud brightening
Black-box engineering	e.g. Air capture (artificial trees)	e.g. Space reflectors, urban albedo enhancement

For all of these reasons, the Royal Society's 2009 report concluded that 'the acceptability of geoengineering will be determined as much by social, legal and political issues as by scientific and technical factors' and recommended 'the development of governance frameworks to guide both the research and development ... and possible deployment' (Shepherd et al., 2009, p. ix). In the following sections, we describe the institutional challenges of geoengineering proposals, the variety of values and interests that account for how institutions position themselves in relation to such proposals, and identify key characteristics of a potential governance framework that could be deployed across a range of political cultures.

Institutional challenges of climate geoengineering

Climate geoengineering presents distinctive institutional challenges for governance because it includes a wide range of possible technologies embodying highly heterogeneous technical practices. Furthermore, they are novel propositions whose full consequences are hard to assess at this early stage.

More broadly, these immature climate geoengineering techniques – and, as Hulme (2007) suggests, climate science itself – fall into a category of what Funtowicz and Ravetz (1990) have characterised as post-normal science (PNS). PNS is policy science when facts are uncertain, values in dispute, stakes high and decisions urgent. Funtowicz and Ravetz suggest that the only way of constructing socially robust science advice under PNS conditions is for scientific institutions to abandon claims to a monopoly of scientific authority, acknowledge problems of uncertainty and quality, and embrace carefully constructed dialogue with wider stakeholders – a process that they call 'extended peer review'.

Indeed, the Royal Society contribution to the geoengineering debate is in the spirit of such an approach. Prior to publication of the 2009 report there was vigorous debate within the Royal Society whether a scientific academy should extend the range of its evaluation beyond narrowly defined issues of physical and biological science to address technology governance issues. The inclusion of social sciences and the law alongside natural sciences and engineering was an example of an institution extending the range of its conventional practice to launch a socially useful debate.

The institutional approach to the challenges of geoengineering adopted by the Climate Geoengineering Governance (CGG) research programme was informed by PNS. It acknowledges the limits of scientific authority where scientific uncertainty and contested values are in play, and the need to take account of multiple possible contexts and consequences of any policy course. In particular this approach enables us to:

- identify the different ways in which geoengineering is framed, in interconnected technical, economic, legal, ethical and social terms;
- explicate the problems of control, or the challenges of controlling the technologies from initial research in computer simulations all the way through

to possible deployment, thinking about such issues as lock-in, path depend-ency and public acceptability; and

- explore governance and regulatory requirements – what kinds of regulatory frameworks might be considered, and why.

We go on to ask what kinds of institutions are suited to confront the issues of safety, efficacy, affordability and public acceptability under various technolo-gical options for pursuing geoengineering. Effective technology governance in the current world has to recognise the interactions between local, national and multinational institutions and the advent of rising powers. The Paris Agreement is itself a piece of institutional innovation that recognises that globally negoti-ated objectives to limit climate change rely on varied national contributions that reflect local geopolitical resources and priorities.

An institutional perspective is also key to understanding policy choice and to assist researchers, policy-makers, the private sector and civil society organisa-tions to define the conditions under which research into, and eventual deploy-ment of, geoengineering measures might be conducted, taking into proper consideration the regulatory and other governance criteria and arrangements which will be required.

How institutions frame geoengineering

The dynamics of emerging technologies can be seen as processes by which a specific set of socio-technical arrangements to pursue a goal are selected, stabil-ised, and institutionalised (Healey, 2014).

Typically, institutional stabilisation allows diverse interests and hetero-geneous technical practices to come together in one sociotechnical image of the future – or 'imaginary' (Jasanoff & Kim, 2009) – that can claim novelty and innovative power and attract political and financial resources. For example, nanotechnology, as a distinctive, discrete framework for understanding techno-logical choices, was stabilised out of molecular engineering, electronic miniaturisation and microbiology, partly through the intellectual entrepre-neurship of Mihael Roco at the US National Science Foundation (see Roco & Bainbridge, 2003).

As we have noted, the term 'geoengineering' is itself very controversial. Under this umbrella term we find a heterogeneous collection of technical prac-tices that have little in common with each other. Technically, the science and engineering of sulphate aerosols have almost nothing in common with tech-niques for carbon dioxide removal from ambient air.

Definitional politics are constantly at work specifying and re-specifying what, at any given moment, might be included under the geoengineering banner. What is on the list will vary over time as a result of the view that proponents of different techniques take of the financial and other resources that defining their research areas as geoengineering may bring. This is balanced against an assess-ment of what additional regulations their field may face. For example, advocates of large-scale afforestation tend to resist forestry being included in geoengineering out of concern that this will place them under additional regulatory burdens. On the other hand, advocates of biochar have lobbied to be included because they view geoengineering as a potential opportunity to develop their favoured technology. In thinking about the governance implications of geoengineering it is important to specify the particular technological practice under consideration. The decision of the US National Research Council (US NRC, 2015a, 2015b) to treat SRM and CDR in two separate reports reflects this continuing definitional politics, seeking to recast geoengineering in general as 'climate intervention' and to domesticate solar radiation management as the less threatening sounding 'albedo modification'.

What all of these geoengineering practices share, for the most part, is their immaturity. The field consists of very early stage technical concepts. While there is much conjecture and speculation about how these might be developed into fully fledged sociotechnical systems, our understanding of what these technologies might be and how they might work in practice at scale remains sketchy and contested (US NRC, 2015a, 2015b). Climate models have been used to simulate the effects of sulphate aerosols, but the reliability of such simulations is contested (e.g. Hulme, 2014a, 2014b).

There are bits of equipment that already exist and could be adapted for incorporation into geoengineering technologies: spray nozzles and aeroplanes, for example. Institutionally, universities, private firms and funding agencies are only just beginning to address whether, and how, to organise geoengineering research. While some international institutions, such as the London Convention and Protocols (LCLP) and the Convention on Biodiversity (CBD) have begun to address issues falling under their purview, there is little consensus about how such technologies should be governed. We have nothing close to a sociotechnical system capable of delivering the kinds of outcomes that geoengineering promises at some point in the future.

Institutional framing and definitional politics

The ambiguity of the term *geoengineering* offers interpretive flexibility for institutions to articulate diverse interests within and across contending framings. It is questionable whether increasing precision in terminology will necessarily facilitate greater clarity in governance discussions or public engagement. The merits of any given form of precision will depend on particular framings. Much ambiguity in this area may thus be irreducible. Hence the challenge lies rather in understanding the wider implications of the range of institutional positions that are at stake.

Markusson (2013; Markusson, Venturini, Laniado, & Kaltenbrunner, 2016) has examined the way in which Wikipedia links represent geoengineering. With the exception of land-based carbon dioxide removal, which clusters with mitigation and adaptation, the analysis shows a clear disconnection between the discourses of geoengineering on Wikipedia and the discourse about mainstream climate change. A possible explanation for this is that several of the land-based technologies for carbon drawdown and sequestration have already been relatively well institutionalised as climate mitigation, whereas sulphate aerosols and ocean fertilisation represent technologies with a less secure role in policy and with less well-developed support bases and funding streams. However, the disconnect between geoengineering and mainstream climate discourse on Wikipedia seems to reflect a similar institutional disconnect in the non-virtual world. This is potentially problematic for geoengineering governance.

Geoengineering has been the subject of two particularly contested framings. One is the climate emergency framing (Heyward & Rayner, 2016) that has previously been deployed in attempts to intensify efforts aimed at conventional greenhouse gas mitigation, but has proven controversial since being appropriated by geoengineering advocates. This framing was quite prominent in the Royal Society report (Shepherd et al., 2009) but subsequently the climate emergency framing has been heavily critiqued, particularly by social scientists who have pointed out that emergencies don't just occur: they are declared. They are political constructions and the very idea of a climate emergency has the potential for the imposition of undemocratic forms of technocratic management of climate and the environment (Hulme, 2014a). It is also impractical (Sillmann, Lenton, Levermann, Ott, Hulme, Benduhn, et al., 2015). When would we know when we were on the verge of a climate emergency? How soon before a climate emergency would we have to know about it, in order to be able to implement any of these technologies to counter it?

A second early framing – the Plan B framing – also proved problematic. The Royal Society report itself does not mention Plan B: that expression is only used in the society president's preface to the report. But it was, nevertheless, a framing that was used prominently in the period from 2010 onwards. This has caused confusion because people have different ideas about what is implied by this. Is Plan B a supplement to Plan A to do a little bit more? Or is Plan B the last resort when Plan A has demonstrably failed? More recently, partly reflecting this ambiguity, the Plan B vocabulary has been dropping out of use.

Institutional trust and liabilities loom larger in geoengineering discourses than in discussions of conventional mitigation and even adaptation. This seems to be because there is a clearer cause-and-effect relationship with a deliberate intervention in the climate, such as geoengineering, than with the accidental disruption of the climate caused by the historic emissions of fossil fuels (although not a relationship so clear that problems of attribution are escaped entirely). Geoengineering also leads us to focus on issues of democracy and consent. Who would have to consent, and how, in order for any agent to engage in either research or the implementation of these technologies at different geographical scales? (Macnaghten & Szerszynski, 2013, Heyward & Rayner, 2015)

Some commentators, (e.g. Hulme, 2014a; Macnaghten & Szerszynski 2013; Cairns, 2014) have suggested that certain forms of geoengineering, specifically sulphate aerosol injection (SAI), are constitutionally incompatible with democracy. The claim that SAI is incompatible, in principle, with democracy is one of three arguments invoked to close down debate and preclude further discussion and research. The other commentators (discussed below) advance the slippery-slope argument, which claims that research into a technology inevitably locks the world into a path to deployment, and the irreducible-ignorance argument that we will never be able to learn enough from sub-scale research to be confident of controllable deployment.

The incompatibility argument hinges on two ideas. First is that a (currently unformed) technology can be characterised by its essential characteristics that make it incompatible with a particular political system. Second, that democracy can similarly be characterised by a singular model of what it is.

Heyward and Rayner (2015) point out that the claim that SAI is inherently undemocratic draws on a particular ideal of democracy modelled on independent communities where citizens are highly engaged and actively involved in their community's decision-making. According to this model, many sociotechnical systems are inherently undemocratic. Examples include energy, telecommunications and air traffic control. In the real world, democracy is highly imperfect and there is every reason to strive to improve it across the board. The question is not how does SAI stack up against an ideal that is seldom, if ever, realised, but whether designing governance arrangements for SAI would make present arrangements better or worse. While there is undoubtedly potential for SAI to exacerbate existing political inequities, it is at least plausible that public discourse and international negotiation around the development and deployment of SRM could actually serve to promote broader democratic goals.

Heyward and Rayner (2015) argue that the eventual shape of the technology is sufficiently underdetermined at this stage as to render such judgements on the implications of SAI for democracy premature at best. If further research on the technologies is required, the key question becomes how to ensure that such research is conducted responsibly and can be halted in the event that society decides that it is moving in an unacceptable direction.

Geoengineering also invites us to consider the issue of humanity's relation to nature and whether it is permissible, even in principle, to deliberately intervene in, or as some would have it, to be 'playing God with the climate' (Fleming, 2007; Hamilton, 2013).

Institutional views of nature

Just as institutions shape definitions of geoengineering, they also shape views of nature that institutions implicitly rely on to inform decision-making. Canadian ecologist C. H. Holling (1986) described the assumptions about nature underlying the seemingly disorganised and contradictory spruce-budworm control strategies practised by foresters in British Columbia.

Holling discerned that there was a consistent pattern in these interventions. His problem was that if the managers were irrational, then there would be no pattern to what they did. If they were all conventionally rational, then they would all do the same thing. Hence, Holling asked, what are the minimal representations of reality that must be ascribed to each managing institution if it is to be considered rational? He found he needed at least three representations, which he called *myths of nature*, each of which could be represented by an icon of a ball in a landscape (Figure 7.1).

The myth[3] of *nature benign* is that the environment is favourable towards humankind. It is a myth of global equilibrium according to which nature renews, replenishes and re-establishes its natural order, regardless of what humans do to their environment. However much the ball is perturbed, the steeply sloping sides of the basin return it to equilibrium. It encourages a trial-and-error approach in the face of uncertainty. Anthropologists have found that this view tends to predominate in competitive-individualist institutional settings where leadership consists of having 'the right stuff'. The institutional bias is one that favours technological innovation and has little time for discourses of constraint (Thompson, 1987; Thompson & Rayner, 1998). Institutions operating with this myth of nature are likely to be receptive to novel technological interventions promising rapid results, such as SAI.

Diametrically opposed to this is the myth of *nature ephemeral*. Far from being stable, nature is seen to be in a precarious and delicate balance. The ball is perched on an upturned bowl and the least perturbation results in a decisive and irreversible change in the state of the system. This myth supports a thoroughly precautionary approach to managing nature that emphasises restraint and has been associated by anthropologists with egalitarian-collectivist institutions that are generally suspicious of techno-fixes.

At first blush, the illustration of the third myth, that of *nature perverse/ tolerant*, might appear to suggest a simple hybrid of the first two. However, it is quite distinctive. Although it acknowledges a certain degree of uncertainty as being inherent in the system, it assumes that scientific management can limit any disorder. The ball will return to equilibrium, provided that measures are taken to ensure that no perturbation is too great. This myth supports neither the unbridled

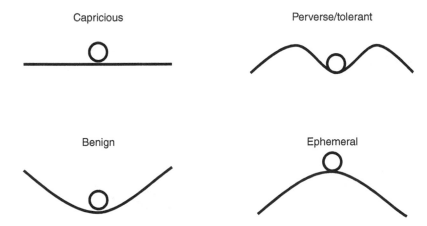

Figure 7.1 Myths of nature.

exploitation of the myth of nature benign nor the cautious, restrictive behaviour of nature ephemeral. Instead, it maps and manages the boundary lines between these two approaches through monitoring, indicators, auditing, and the construction of elaborate technical assessments (Rayner, 2003), typically demanded by predominantly hierarchical institutions, especially when making large infrastructure investments and long-term investments, such as those envisaged for geoengineering.

To these three nature myths, anthropologist Michael Thompson (1987) added a fourth: that of *nature capricious*, represented by a ball on a flat surface, liable to move unpredictably in response to any perturbation. The myth of nature capricious is associated with a fatalistic world outlook that does not actively engage in managing nature, which is, in principle, unmanageable. Hence, this institutional setting engenders indifference and scepticism towards any kind of climate policy.

It is instructive to apply the same iconography to perceptions of the economy (see Figure 7.2). While institutional discourses that frame nature as capricious or perverse/tolerant tend to see the economy in the same terms, the icons for the economy are the inverse of those representing nature in the other two perspectives. That is to say, institutions that view nature as benign and forgiving tend to be most worried about the impact of environmental policies on the economy. Those of an egalitarian orientation, who fear that nature is ephemeral, tend to be convinced that greening the economy will not only assure the security of ecosystems but will also improve economic performance. This situation was clearly visible in the contrasting approaches taken by George W. Bush and Al Gore on climate change. Bush, reflecting a competitive-individualist orientation, repeatedly warned that the US would not compromise its economic position and the American way of life by implementing policies to reduce carbon dioxide emissions from energy production and use. In contrast, in 2004 Gore co-founded Generation Investment Management based on his commitment to the idea that

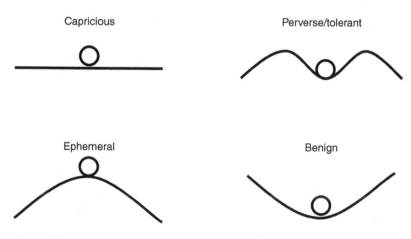

Figure 7.2 Myths of the economy.

the same kinds of environmental policies deplored by Bush would open up new economic growth opportunities for green investment.

These different institutional orientations have implications for the acceptability of climate geoengineering. The motivation for climate policy, to date, has emerged as a discourse of constraint (emissions reductions) and an alignment between egalitarian and hierarchical views in opposition to competitive-individualist scepticism, if not about climate change itself, at least about its allegedly catastrophic consequences. The alignment between egalitarian and hierarchical perspectives has sought to accommodate the competitive-individualist preference for market solutions through proposals for carbon pricing that have largely failed to gain traction. However, in the competitive-individualist institutional setting, geoengineering holds out the prospect of bold human intervention to redress any possible climate disruption. This suggests the possibility of a realignment of climate policy should the hierarchical institutions become convinced either that geoengineering is necessary to avoid imminent catastrophe (i.e. that the ball in their myth of nature is close to a tipping point) or that it offers an economically less disruptive path to climate stabilisation than relying on radical emissions cuts. In either case, hierarchical institutional actors could find themselves relying more heavily on geoengineering strategies at the expense of the egalitarian preference for efficiency and decentralised renewables.

Thus, the prospect of a political realignment of hierarchies with competitive-individualist orientations based on geoengineering is profoundly threatening to egalitarian institutions, such as activist groups, that currently rely on the threat of climate chaos to maintain their influence with governmental hierarchies (Heyward & Rayner, 2016).

It has also been suggested (Rayner, 1991; O'Riordan, Cooper, Jordan, Rayner, Richards, Runci, et al., 1998) that each type of institution favours different kinds of policy instruments. Competitive-individualist institutions favour carrots over sticks and prefer fiscal incentives and price mechanisms over regulations; they particularly favour market-type instruments that leave discretion with firms and individuals. Egalitarian institutions generally lean towards sticks in the form of unambiguous regulations, which allow minimum discretion in implementation, as the only way to keep both the regulated and the regulators honest. Hierarchical institutions tend to be pragmatic about deploying both carrots and sticks, largely at the discretion of the regulator. All three approaches can be identified in the discourses around climate geoengineering, adding to the complexity of the governance challenges.

Security concerns

Hierarchical institutions are also particularly concerned with maintaining domestic public order and international peace and security. Geoengineering proposals have given rise to concerns that they could be destabilising of international security (Nightingale & Cairns, 2014). For example, some commentators have expressed concern that geoengineering technologies might be weaponised (Robock, 2008; Fleming, 2010). However, there are reasons to be sceptical about such claims. The principal military applications for weather and climate control appear to be terrain denial and demoralisation of civilian populations, but there are easier, cheaper and more targetable options already available for these functions.

Furthermore, existing international legal institutions would constrain such applications. Any state's use of geoengineering in a fashion that deliberately or accidentally injures another state would be a violation of existing customary international law. Any explicitly military application would violate an existing (if hitherto inactive) treaty: the United Nations Convention on the Prohibition of Military or Any Other Hostile Use of Environmental Modification Techniques (ENMOD) already prohibits environmental warfare.

But that does not mean to say that these technologies are without any security implications at all. The indirect threat of cross-border impacts is a very real one. Political actions are based on perceptions. Geoengineering activity by one state could become a lightning rod for conflict with neighbours, especially where existing tensions are high. For example, it is easy to imagine that had India conducted some sulphate aerosol injection experiments immediately before the 2010 floods in Pakistan, members of the Pakistani public, and certainly Pakistani politicians, would have held India responsible for the damage that had been caused by the floods.

The political risks of this kind of damage claim over geoengineering are deepened by the potential involvement of the military with some kinds of geoengineering deployment. The implementation of technologies like SAI would be carried out by, or in conjunction with, the institutional web of the military-industrial complex. In Europe companies like BAE Systems, EADS or Finmeccanica have particular skills and capacities that would make them likely equipment suppliers. The likes of General Dynamics, Lockheed Martin or Halliburton would be likely suppliers or even contractors implementing any US-based programme. Participants in a CGG workshop in China left us in no doubt that if that country were to adopt any kind of SRM, implementation would fall to the People's Liberation Army, which currently conducts the vast array of weather modification activities in that country.

In the US, the national security linkage would likely occur not only because the military would have the relevant technical capacities, but because the risks of a rapid temperature spike caused by a sudden cessation of SAI would mean that such a project would be considered part of the country's critical infrastructure from the very start: thus subject to the closest national control (Nightingale & Cairns, 2014). The network of defence and security organisations that would become involved might increase institutional lock-in, making it more difficult to modify or withdraw SAI once deployed. This possibility makes it more desirable to seek governance mechanisms by which geoengineering initiatives can be agreed, and any disputes that might arise over their impacts can be settled.

National and international law and governance

Prompted by this concern, and in line with the promptings of the Royal Society Report, Armeni and Redgwell (2015a, 2015b) analysed both international and domestic legal arrangements to assess their potential to respond to existing imaginaries of geoengineering and what new provisions might be necessary.

At the international level, they highlighted the absence of any existing international treaty broadly applicable to geoengineering, except for the 1977 ENMOD treaty already referred to. This, however, assumes the intention to cause harm. General legal principles applying to the conduct of states would require, in any case, that they exercise due diligence to ensure that nothing they did would cause significant harm across boundaries with other states, or the high seas or air space above. Otherwise, no single institution or instrument ticks all boxes as a one-stop shop for the international governance of geoengineering. The Convention on Biological Diversity (CBD) comes closest, but lacks robust enforcement mechanisms, and, given its biodiversity focus, is not an instrument of general application. At the level of specific technologies, the London Convention/London Protocol (LC/LP), originally designed to regulate ocean dumping of waste, has shown itself to be a flexible institution for dealing with ocean iron fertilisation: initially promulgating non-binding guidelines and then progressing to binding arrangements. It could provide a useful template in designing regulation of other specific technologies.

The atmosphere – the domain where solar radiation management (SRM) technologies would be deployed – constitutes a clear case of a gap requiring new institutional machinery. While it is generally acknowledged that state control extends to the air space above national territory, there is no direct national control of air space in areas beyond national jurisdiction, nor any single instrument that regulates the introduction of substances, e.g. sulphate aerosols, into the stratosphere.

CGG researchers Armeni and Redgwell (2015c, 2015d) also examined the legal institutional landscapes of three exemplary jurisdictions in which geoengineering has been discussed – the USA, Germany and the UK – to explore the potential role of national laws and regulations. There are national particularities. For example, freedom of scientific research is specifically protected under German Basic Law. But there are also commonalities such as that the heterogeneity and scope of geoengineering proposals means that no single regulatory institution or government department in any country is likely to have a full mandate to regulate research or implementation, even within its own national boundaries. Indeed, the legal and regulatory implications of geoengineering will spill over the boundaries of environmental or technological controls as they are likely to have implications for trade, food supply, intellectual property rights and, as already discussed, national security; collaboration across these domains will be required at both international and national levels as appropriate.

Institutionalisation of regulation could take three (not mutually exclusive) forms: adapting existing provisions; bespoke coverage of gaps in those provisions; and developing general principles of governance. In developing any institutional framework, it is necessary to be aware of regulatory intention; is the law seeking to keep options open, or to limit potential risks?

At the international level, there is some scope for adaption of existing institutional machinery as with LC/LP, but the regulation of what is done within the atmosphere constitutes a real gap requiring new provisions. This might be considered a priority given the attention currently being given to SRM technologies.

In summary, there is no one existing institution with the scope and competence to cover the whole range of either international or national governance requirements for geoengineering. Indeed, there is no one-size-fits-all approach to geoengineering governance and, given the heterogeneous nature of the proposals, such an approach seems undesirable compared to a bootstrapping or 'experimentalist-governance' (Armeni, 2015) approach of developing the institutional frameworks in closely coupled feedback loops with the development of the technologies themselves (Rayner et al., 2013).

Voluntary, bottom-up arrangements have a contribution to make, but require a minimum of an oversight organisation to keep track of what each country is doing on regulation and attribution, and to guard against regulatory fragmentation. Finally, any machinery requires flexibility to be able to cope with changes in the technologies themselves or with the capacities to monitor and attribute their impacts. As argued above, the technologies remain under-specified. Before their implementation can be regulated, they will require extensive research and development. What are the appropriate institutional arrangements and safeguards for research?

Regulating geoengineering research

The first problem facing any effort to institutionalise the regulation of geoengineering research is distinguishing geoengineering research from basic science. For example, a geoengineering experiment may be functionally indistinguishable from a basic science experiment to explore how droplets form around particles in the stratosphere or to explore marine cloud formation. Opinion remains divided as to whether intentions matter. Regulators will have to decide whether researchers should only be subject to controls if they say 'I'm doing this for geoengineering', but not if they say they are doing basic science?

A similar problem arises from the view that solar radiation management, and particularly SAI, could only be fully tested at large scale. Where is the cut-off point, if any, between research and implementation? What are the institutional implications of such a distinction? These are highly contested areas. They have not been resolved. They continue to be subject to contestation.

The notion of the 'slippery slope' has particularly concerned those who advocate a research moratorium or a ban and who worry that once we take the first step in developing these technologies then implementation becomes inevitable (Morrow, Kopp, & Oppenheimer, 2009; ETC Group, 2010; Hamilton, 2013; Hulme, 2014a). Counter to such concerns, billions of dollars were invested in developing fast breeder reactors around the world and no commercial breeder reactors exist today. So, the notion that a slippery slope is inevitable does not appear to be verified. Similar to the concern about the slippery slope are notions of path dependence and lock-in.

Although all programmes of innovation need a degree of sociotechnical commitment and, therefore, lock-in to succeed, society needs to avoid lock-in to inappropriate or unacceptable technology and to inappropriate institutional architecture to manage and regulate it (Wong & Markusson, 2015).

Geoengineering would seem to be susceptible to institutional lock-in as many of the proposed technologies (e.g. direct air capture), would be dependent on the existence of a highly capital-intensive physical infrastructure, with large sunk costs creating vested interests in keeping facilities operational, and cognitive lock-in whereby early institutional framing of the problem shapes the outcomes of technology assessments (Cairns, 2014).

Perspectives on institutional architecture for geoengineering governance reflect a persistent tension between advocates of top-down anticipatory structures – those who feel that there ought to be some kind of global treaty in place, even before research is conducted (Bodle & Oberthuer, 2014) – and those who favour a bottom-up, emergent approach to geoengineering governance (Long, Loy, & Morgan, 2015). As indicated earlier, there has been some support for the idea of the Convention on Biodiversity (CBD) being the overall top-down body, but others have questioned whether that is actually a suitable role for the CBD and whether it has the potential to undertake it.

There is also a range of dilemmas for the control of research: which institutions should govern and regulate it? As well as the idea that there ought to be some kind of overarching international institutional framework, different institutional arrangements have been proposed, ranging from self-governance by scientists, through national government regulation, to the idea that scientific academies or funding bodies should regulate it. Since the latter have the prime responsibility to decide the allocation of research resources they often have the best technical information about what is proposed.

But there is the question how much of a comprehensive institutional architecture is required even before embarking on research. Should there even be research (Rayner, 2015)? On this issue, British social geographer Mike Hulme (2014a) has advanced the 'irreducible ignorance' argument, referred to above, that we simply cannot know enough about how SRM would behave to justify doing it: he sees the climate models that are used to assess its impacts as mere 'calculative cartoons' relative to the reality that we would have to deal with. Furthermore, he argues that SRM is morally reprehensible, so our ignorance will save us from folly. By way of contrast, and reflecting a quite different institutional background, David Keith (2013), a US engineer, has set out a programme by which he believes that ignorance can be reduced through carefully constructed, carefully staged, escalating research, and argues, contrary to Hulme, that it would be folly to remain ignorant of our ignorance. Is it possible to improve our knowledge of the potential performance of SRM in an environmentally and socially responsible manner?

An institutional architecture for research governance

The Oxford Principles for Geoengineering Governance were initially proposed to the UK House of Commons Science and Technology Committee, as guidelines for designing institutional arrangements for research governance and eventual possible deployment (Rayner, Heyward, Kruger, Pidgeon, Redgwell, & Savulescu, 2013). They are that:

1   Geoengineering should be regulated as a public good and in the public interest;
2   There should be participation in decision-making at the appropriate level, depending on whether the activity of concern is computer-based research, outdoor research limited to a locality, or to be conducted on a national or international scale;
3   There should be full disclosure of geo-engineering research results and open publication of all results. (This was thought to be important in the light of experience in the pharmaceutical industry, which has tended not to publish unfavourable clinical trial results.);
4   There should be independent assessments of the research and particularly of impacts;
5   The governance arrangements need to be clear prior to any research or deployment.

These general principles could be applied to almost any potentially controversial area of technology innovation, but need to be made specific in the instance of geoengineering. To this end, the authors of the Oxford Principles suggest combining the idea of technology-specific protocols with stage-gates. This would require that research be structured as a series of phases. At the start of each one there would have to be an explicitly articulated plan showing how the Oxford Principles were being addressed. Once each stage of research is completed, before continuing to a further stage, researchers and funding bodies would assess how the principles had actually been implemented in the previous stage and then specify how they would be addressed in the next stage. This kind of institutional arrangement could enable the pursuit of better knowledge and understanding while offering some reasonable assurance that society retains both the right and the institutional capacity to make judgements to close down research, should it deem it appropriate to do so.

Some researchers are currently proposing to conduct SRM experiments in the ambient environment. While these would be on a scale that would appear to present no immediate physical danger, this is arguably the most controversial of geoengineering proposals. Since there is, as yet, no accepted boundary between SRM research and implementation, it is important that early experiments establish a clear pattern of environmental and social responsibility from the beginning.

Clearly, governments, research funding organisations and scientific and professional bodies require the establishment of an open and transparent review process that ensures that any such experiments have the necessary social licence to operate. At a minimum, such a review process should involve prior open publication of research plans, in order to assure the integrity of the process, with independent evaluation of all existing evidence, plans and results, and should actively seek public participation. Such a process will promote the scientific integrity and public legitimacy of any such experimental work. The approach proposed by the authors of the Oxford Principles does not make any assumptions about the specific institutions that would implement such arrangements. These could vary according to the scale and nature of the proposed actions, legal and political jurisdictions, and stage of development in the research, development, demonstration and deployment process. It therefore exemplifies the bootstrapping approach, described above, by which institutional arrangements develop one step ahead of the techniques, rather than, at the outset, attempting to design a comprehensive institutional architecture that might prove either excessively restrictive or inadequately permissive in the light of subsequent developments.

Discussion and conclusions.

From a technical viewpoint, the obvious question about climate geoengineering is what can it do for climate? Answering that question will require extensive research and development, which will require institutional arrangements to promote, perform and regulate the research process. Social scientists have begun to address the question whether the CDR technologies could feasibly be scaled up in a timely manner to the level necessary to have the impact that is implicit in the Paris Agreement. They also recognise that while solar radiation management promises high leverage and fast-acting impacts it would be very difficult to govern. These are essentially institutional questions.

But social scientists also want to ask: what can geoengineering, particularly geoengineering discourses, do for society? At the moment, unlike the development of cases like the nanotechnology and GM disputes, we are having the discussion about values before the technology is developed.

At this stage, there is little science to hide those values behind. So, geoengineering provides opportunities to explore things like the way we think about nature, what we think is the good society, and the role of technology in our lives. What are the implications for social justice of technological interventions? And what can geoengineering teach us about the governance of other emerging global technologies?

In debates on emerging new technologies, discussions about the science typically displace explicit discussion of the institutional values that may lie behind them. Controversy about GM crops and nanomaterials, for example, only arose after the technologies had been developed and already achieved a significant level of adoption (at least in some parts of the world). In these cases, there was already a significant body of scientific knowledge that provided a surrogate arena for contesting values. Because geoengineering science is currently underdeveloped, the values and underlying myths of nature and the economy informing geoengineering debates are unusually prominent. The value differences are clear, and there are no strong science claims to hide behind in articulating those values (Sarewitz, 2004). Quite the contrary, everyone seems to acknowledge that we currently know very little. Once research gets under way (if it does), then the debates will likely change, and explicitly begin to focus on what is being learned by the research, and what it tells us. And once that happens, debates over the science will displace the debate over the values (Rayner, 2015). Such a shift would of course suggest a gap in our institutional arsenal: a mechanism by which there can be more continual stakeholder and public participation in devising robust sociotechnical scenarios – public imaginaries – in strategic areas of policy choice.

More broadly we conclude that there is a pressing need to consider a flexible and responsive institutional design that can evolve in a timely manner to ensure that any development of geoengineering technologies is conducted responsibly and in the broader context of emissions mitigation and impacts adaptation measures. Given the environmental, technical and social complexities of geoengineering and the diverse range of institutional preferences for governance and regulation discussed in this chapter, we argue for a pluralist approach to institutions embodying, in the terms of O'Riordan et al. quoted at the outset, formal organisational structures and informal networks of communication, values, norms and expectations that 'range from the formal deliberating bodies engaged in treaty making to the informal liaisons among policy analysts and policy executives, regulatory agencies, and the day-to-day activities of billions of people'.

Notes

1 The research reported in this chapter (including CGG Working Papers and several of the open-literature publications cited herein) was conducted under the Climate Geoengineering Governance Programme funded by the Economic and Social Research Council (ESRC) and the Arts and Humanities Research Council (AHRC) – grant ES/J007730/1. The Principal Investigator was Steve Rayner. Peter Healey was Programme Manager. See http://geoengineeringgovernanceresearch.org.
2 *Die Zeit* quotes a protester saying that the protest movement against carbon storage played into existing German environmental narratives by consciously adopting the same German term 'Endlager' for the final storage of both nuclear waste and carbon. www.zeit.de/2010/47/U-CSS-Widerstand/komplettansicht
3 Holling follows the anthropological usage of 'myth' as a foundational narrative of a society or institution rather than the more popular understanding of a widely believed falsehood.

References

Anderson, K. (2015). Duality in climate science. *Nature Geoscience, 8*, 898–900.
Armeni, C. (2015). Global Experimentalist Governance, International Law and Climate Change Technologies. *International and Comparative Law Quarterly, 64*(4), 875–904.
Armeni, C., & Redgwell, C. (2015a). *International legal and regulatory issues of climate geoengineering governance: rethinking the approach.* CGG Working Paper 21. Oxford: Institute for Science, Innovation and Society, University of Oxford.

Armeni, C., & Redgwell, C. (2015b). *Assessment of International Treaties Applicable, or At-Least Adaptable, to Geoengineering-Related Activities Through Indicators.* CGG Working Paper 22. Oxford: Institute for Science, Innovation and Society, University of Oxford.

Armeni, C., & Redgwell, C. (2015c). *Geoengineering Under National Law: A Case Study of the United Kingdom.* CGG Working Paper 23. Oxford: Institute for Science, Innovation and Society, University of Oxford.

Armeni, C., & Redgwell, C. (2015d). *Geoengineering Under National Law: A Case Study of Germany.* CGG Working Paper 24. Oxford: Institute for Science, Innovation and Society, University of Oxford.

Bodle, R., & Oberthuer, S. (2014). *Options and Proposals for the International Governance of Geoengineering.* Dessau: German Federal Environment Agency.

Cairns, R. C. (2014). Climate geoengineering: issues of path-dependence and socio-technical lock-in. *Wiley Interdisciplinary Reviews: Climate Change, 5*(5), 649–661.

ETC Group. (2010). *Geopiracy: The case against geoengineering.* ETC Group Communiqué, 103.

Fleming, J. R. (2007). The Climate Engineers: Playing god to save the planet. *Wilson Quarterly, 31*(2), 46–60.

Fleming, J. R. (2010). *Fixing the Sky: The Checkered History of Weather and Climate Control.* New York: Columbia University Press.

Funtowicz, S., & Ravetz, J. R. (1990). *Uncertainty and quality in science for policy.* Dordrecht: Kluwer Academic Publishers.

Fuss, Sabine, Canadell, Josep G., Peters, Glen P., Tavoni, Massimo, Andrew, Robbie M., Ciais, Philippe, Jackson, Robert B., Jones, Chris D., Kraxner, Florian, Nakicenovic, Nebosja, Le Quéré, Corinne, Raupach, Michael R., Sharifi, Ayyoob, Smith, Pete, & Yamagata, Yoshiki (2014). Betting on Negative Emissions, *Nature Climate Change, 4*, 850–853.

Gallie, W. B. (1995). Essentially contested concepts. *Proceedings of the Aristotelian Society, 56*, 167–98.

Giddens, A. (1986). *Sociology: A Brief but Critical Introduction.* London: Macmillan.

Gould, S. J. (1991). Institution. In V. Bogdanor (Ed.) *The Blackwell Encyclopedia of Social Science.* Oxford: Basil Blackwell.

Hamilton, C. (2013). *Earthmasters: Playing God with the Climate.* London: Allen & Unwin.

Healey, P. (2014). *The Stabilisation of Geoengineering: Stabilising the Inherently Unstable?* CGG Working Paper 15. Oxford: Institute for Science, Innovation and Society, University of Oxford.

Heyward, C., & Rayner, S. (2015). Uneasy Expertise: Geoengineering, Social Science, and Democracy in the Anthropocene. In M. Heazle & J. Kane (Eds.) *Policy Legitimacy, Science and Political Authority: Knowledge and Action in Liberal Democracies.* Abingdon: Routledge Earthscan.

Heyward, C., & Rayner, S. (2016). Apocalypse Nicked: Stolen Rhetoric in Early Geoengineering Advocacy. In S. Crate & M. Nuttall (Eds.) *Anthropology and Climate Change: From Actions to Transformations.* Walnut Creek: West Coast Press.

Holling, C. S. (1986). Resilience of terrestrial ecosystems; local surprise and global change. In W. C. Clark and R. E. Munn (Eds.) *Sustainable Development of the Biosphere.* Cambridge: Cambridge University Press.

Hulme, M. (2007, March 14). The appliance of science. *Guardian.*

Hulme, M. (2014a). *Can Science Fix Climate Change? A Case Against Climate Engineering.* Cambridge: Polity Press.

Hulme, M. (2014b). Attributing weather extremes to 'climate change': a review. *Progress in Physical Geography, 38*(4), 499–511.

Jasanoff, S., & Kim, S.-H. (2009). Containing the Atom: Sociotechnical Imaginaries and Nuclear Power in the United States and South Korea, *Minerva, 47*, 119–146.

Keith, D. (2013). *A Case for Climate Engineering.* Cambridge MA: MIT Press

Long, J. C. S., Loy, F., & Morgan, M. G. (2015). Policy: Start research on climate engineering. *Nature, 518*, 29–31.

Macnaghten, P. & Szerszynski, B. (2013). Living the global social experiment: An analysis of public discourse on solar radiation management and its implications for governance. *Global Environmental Change, 23*(2), 465–474.

Markusson, N. (2013). *Tensions in framings of geoengineering: Constitutive diversity and ambivalence.* CGG Working Paper 3. Oxford: Institute for Science, Innovation and Society, University of Oxford.

Markusson, N., & Wong, P-H. (2015). *Geoengineering Governance, the Linear Model of Innovation, and the Accompanying Geoengineering Approach.* CGG Working Paper 20. Oxford: Institute for Science, Innovation and Society, University of Oxford.

Markusson, N., Venturini, T., Laniado, D., & Kaltenbrunner, A. (2016). Contrasting medium and genre on Wikipedia to open up the dominating definition and classification of geoengineering. *Big Data & Society, 3*(2), 1–17.

Morrow, D. R., Kopp, R. E., & Oppenheimer M. (2009). Toward ethical norms and institutions for climate engineering research. *Environmental Research Letters, 4*(4), 045106.

Nightingale, P., & Cairns, R. (2014). *The security implications of geoengineering: Blame, imposed agreement and the security of critical infrastructure.* CGG Working Paper 18. Oxford: Institute for Science, Innovation and Society, University of Oxford.

O. Riordan, T., Cooper, C. L., Jordan, A., Rayner, S., Richards, K. R., Runci, P., & Yoffe S. (1998). Institutional Frameworks for Political Action. In S. Rayner & E. L. Malone (Eds.), *Human Choice and Climate Change: An International Assessment, Volume 1, The Societal Framework.* Columbus: Battelle Press.

Rau, G. H., & Greene, C. H. (2015). Emissions reduction is not enough. Letter to *Science, 349*(6255), 1459.

Rayner, S. (1991). A Cultural Perspective on the Structure and Implementation of Global Environmental Agreements. *Evaluation Review, 15*(1), 75–102.

Rayner, S. (2003). Democracy in the Age of Assessment: Reflections on the Roles of Expertise and Democracy in Public-Sector Decision Making. *Science and Public Policy*, 30(3), 163–170.

Rayner, S. (2015). To Know or Not to Know? A Note on Ignorance as a Rhetorical Resource in Geoengineering Debates. In M. Gross & L. McGoey (Eds.), *International Handbook of Ignorance Studies.* Abingdon: Routledge.

Rayner, S. (2016). Editorial: What Would Evans-Pritchard Have Made of Two Degrees? *Anthropology Today, 32*(4), 1–2.

Rayner, S., Heyward, C., Kruger, T., Pidgeon, N., Redgwell, C., & Savulescu J. (2013). The Oxford Principles for Geoengineering Governance. *Climatic Change, 121*(3), 499–512.

Ricke, K. L., Morgan, M. G., & Allen, M. R. (2010). Regional climate response to solar-radiation management. *Nature Geoscience, 3*, 537–541.

Robock, A. (2008). 20 Reasons why Geoengineering may be a Bad Idea. *Bulletin of the Atomic Scientists, 64*(2), 14–18.

Robock, A. (2012). Will geoengineering with Solar Radiation Management Ever Be Used? *Ethics, Policy and Environment, 15*(2), 202–205.

Roco, M. C., & Bainbridge, W. S. (2003). *Converging technologies for improving human performance: Nanotechnology, biotechnology, information technology and cognitive science*. Dordrecht: Kluwer Academic Publishers.

Sarewitz, D. (2004). How Science Makes Environmental Controversies Worse. *Environmental Science and Policy, 7*, 385–403.

Shepherd, J., Caldeira, K., Cox, P., Haigh, J., Keith, D., Launder, B., Mace, G., McKerron, G., Pyle, J., Rayner, S., Redgwell, C., & Watson A. (2009). *Geoengineering the Climate: Science, Governance and Uncertainty*. London: The Royal Society.

Sillmann, J., Lenton, T. M., Levermann, A., Ott, K., Hulme, M., Benduhn, F., & Horton J. B. (2015). Climate emergencies do not justify engineering the climate. *Nature Climate Change, 5*(4), 290–292.

Smith, R. M. (1988). Political Jurisprudence, the 'New Institutionalism' and the Future of Public Law. *American Political Science Review, 82*(1), 89–108.

Thompson, M. (1987). Welche Gesellschaftklassen sind potent genung, anderen ihre Zukunft aufzuoktroyieren? In Burchardt, L. (Ed.), *Design der Zukunft*. Cologne: Dumont.

Thompson, M. (2008) *Organising and Disorganising. A Dynamic and Non-Linear Theory of Institutional Emergence and Its Implication*. Axminster: Triarchy Press.

Thompson, M., & Rayner, S. (1998). Cultural Discourses. In S. Rayner & E. L. Malone (Eds.), *Human Choice and Climate Change: An International Assessment Volume I, The Societal Framework*. Columbus: Battelle Press.

Tollefson, J. (2015). The 2°C dream. *Nature, 527*, 436–438.

UNEP (2016). *The Emissions Gap Report* 2016. Nairobi: United Nations Environment Programme.

US NRC (2015a). *Climate Intervention: Reflecting Sunlight to Cool Earth*. Washington DC: National Academies Press.

US NRC (2015b). *Climate Intervention: Carbon Dioxide Removal and Reliable Sequestration*. Washington DC: National Academies Press.

Wong, P.-H. & Markusson, N. (2015). *Geoengineering Governance, the Linear Model of Innovation, and the Accompanying Geoengineering Approach*. CGG Working Paper 20. Institute for Science, Innovation and Society, University of Oxford.

8 Post-Paris long-term climate capacity

The role of universities

Naznin Nasir, Riadadh Hossain, and Saleemul Huq

Introduction

Global communities are increasingly recognising that limiting greenhouse gas emissions and adapting to the changing climate is fundamental for achieving a low-carbon, climate resilient future. Appropriate and adequate capacity remains integral for implementing climate change mitigation and adaptation initiatives. Unfortunately, not all countries have similar capacities to address climate change and, ironically, often countries with grave deficiencies in their capacity – such as the least-developed countries (LDCs) and the small island developing states (SIDS) – tend to be the most climate vulnerable (Hoffmeister, Averill, & Huq, 2016). These countries require skilled individuals, organisations, and political and economic systems to ensure they are able to plan and implement actions for achieving national and global climate goals (Chen & He, 2013). As such, capacity building is fundamental for achieving the goals established in the Paris Agreement for combating climate change. Under the Paris Agreement, capacity building provisions such as the decision to create the Paris Committee on Capacity Building (PCCB) (Article 11); a Capacity Building Initiative for Transparency (CBIT) (Article 13); and to promote education, training and public awareness (Article 12) have been made to facilitate global climate change capacity development (Khan, Sagar, Huq, & Thiam, 2016).

These provisions made under the agreement also intend to change the existing short-term consultant-driven capacity-building model to a long-term and sustainable model. Universities, which for centuries have been revered as hubs of innovation and learning, have for the most part been disengaged from global climate capacity-building initiatives. However, many university-based researchers, educators and students are already deeply involved in producing, communicating and learning climate knowledge and skills (Hoffmeister et al., 2016). Thus, it can be perceived that universities are crucial agents for creating lasting climate capacity-building programmes.

This chapter seeks to highlight the role of universities in achieving capacity-building goals following the vision for long-lasting capacity-building initiatives laid out in the Paris Agreement. In doing so, it highlights the challenges and shortcomings of existing capacity-building programmes and also describes how the issue of capacity building has historically been dealt with under the UN Framework Convention on Climate Change (UNFCCC). The chapter also provides some suggestions as to how universities can play a greater role in minimising capacity gaps between the Northern and Southern institutes and identifies some potential challenges within university systems that will have to be addressed to ensure that they can play an effective role in building climate capacity.

Limitations of current capacity-building practices: the need for a paradigm shift

Article 11 of the Paris Agreement called for sustainable climate capacity building and accordingly the Paris Committee on Capacity Building has been created. PCCB challenges the old paradigm of capacity building in many ways and demands a new way of building capacity. As yet, capacity building on climate change has mostly been conducted on a piecemeal basis. Developed countries usually have their own capacity-building programmes and agendas and they choose which developing countries they will support. There is no effective coordination mechanism or authority to ensure effective and sustainable capacity building. Developed countries allocate funds for projects or programmes aimed at climate change capacity building in their target countries and allocate the fund to their Official Development Assistance (ODA) agencies. The designated development agencies often assign private consultancy companies from their own country that send consultants to the designated country mostly for providing short-term assistance which, in most cases, takes the form of workshops and presentations. The consultants usually do not know the local language, and so conduct their workshops in their own language (which of course is often a second language to the audience, whose capacities they are trying to build (Huq & Nasir, 2016)). They often do not know the country context they are working in. The designated consultants make a few more visits, write a report and it is assumed that 'capacity building has happened' (Huq & Nasir, 2016).

A major problem with this existing model is that this 'fly in and fly out' approach leaves little to no long-term capacity in the targeted developing country (Huq, 2016a). A few key people may become better versed, but institutions themselves are not capacitated. Additionally, a big chunk of the budget that has been spent on developed countries' own consulting companies is then often passed off as 'obligated Overseas Development Assistance' or climate finance. This vicious cycle may only help to achieve short-term political objectives of the Global North. In order to ensure effective and sustainable capacity building, stronger commitments are therefore essential. Negotiators and experts from the Global South believe that sustainability of capacity building is the key prerequisite for ensuring effective usage of allocated fund. During COP21 this argument became stronger when negotiators from the developing countries said that current capacity-building practices were inefficient, costly and unsustainable.

Climate change capacity-building programmes are expected to be long-term and flexible, and also to be able to adapt to changing conditions and complex challenges, which demands sustainable financial resources as well. Unfortunately, in many cases capacity-building programmes have been short-term and mostly consultant-driven. For example, many of the capacity-building projects or programmes were merely designed to prepare reports and documents (e.g. the INDCs and NAPs) required by the UN climate process. Unfortunately, simply preparing documents to submit to the UN climate body does not count as long-term capacity building. These documents often do not have resonance in national planning, and these programmes do not create any sense of national ownership either, as they are often prepared by hired consultants. Long-term capacity building, as opposed to short-term capacity building, would invest in national systems. Chen and He (2013) argued that capacity-building programmes are more likely to be successful where they are country driven, and when there is a sufficient degree of country ownership. Thus, more pragmatic and ideal way of doing things would be to engage institutions within the countries themselves for driving capacity building.

Another limiting aspect of the current capacity-building model is the one-way learning process. The presumption in the current model is that the international consultant has to teach the people of the developing country, and does not have much to learn from them as he or she is already an 'expert'. This paradigm of one-way learning needs to be altered and the model should promote two-way learning so that the relationships developed over time are those of mutual learning and knowledge exchange between equals rather than the 'expert teaching the student'.

The Global North has years of experience helping developing countries, primarily in acquiring climate mitigation knowledge. However, the Global South has also developed the expertise of adapting to the adverse effects of climate change; lessons from which need to be disseminated worldwide. Consequently, capacity building need not be a one-directional road from North to South but should rather facilitate North to South, South to North and South to South knowledge exchange programmes as well. For instance, when it comes to climate change adaptation, countries may learn more from developing countries that have greater practical hands-on experience of adapting to climate change.

The UNFCCC reported 681 capacity-building activities undertaken in 2015, by 16 international institutions – an increase of over 80% from 2012 (UNFCCC, 2016a). Over the 2009–2015 period, support for mitigation-related capacity building was reported as increasing from US$15.75 million to US$321.16 million. While most capacity-building programmes funded by developed countries focus on how to reduce greenhouse gas emissions through mitigation, most of the demand from developing countries, and especially from the poorest ones, is for capacity building on adapting to climate change. Yet less money is going towards capacity building for adaptation. Thus, the current funding paradigm for capacity-building programmes also needs to shift so that country priorities would be given preference over donor choice.

Capacity Building under UNFCCC

History of capacity building before Paris Agreement

Capacity building has been part of international climate negotiations under the United Nations Framework Convention on Climate Change (UNFCCC) since its inception in 1992. Article 6 of the Convention recognises the importance of promoting education, training and public awareness on climate change, both at the global and local level and across different segments of the population (UNFCCC, 1992). However, the issue of capacity building has historically been one of the least contested issues at the climate negotiations until COP21. In fact, there has been so little controversy around the issue that developed countries often assigned their least experienced negotiators in the discussions (Hoffmeister et al., 2016)

In the 7th Conference of Parties (COP7) held in 2001, the Marrakech Accords established a capacity-building framework which still guides all capacity-building activities under the UNFCCC. The guiding principles of the framework highlight the need for capacity building to be:

- Country-driven and based on the priorities of developing countries;
- A continuous, progressive and iterative process;
- Undertaken in an effective, efficient, integrated and programmatic manner;
- Cognisant of special circumstances of LDCs and SIDs;
- Conducive to 'learning by doing';
- And to rely on and mobilise existing national, sub-regional and regional institutions and the private sector and build on existing processes and inherent capacities.

This framework has been reviewed on a regular basis, with the first review conducted during COP10 in 2004 and the second review during COP13 in 2007. Following the reviews, a number of gaps were identified and several elements were recognised for consideration for future implementation (UNFCCC, 2016b).

Figure 8.1 depicts how the agenda of capacity building has evolved over the years in the international climate negotiation landscape.

The Paris Agreement sets a roadmap on capacity building

The agenda of capacity building gradually gained momentum in the international climate negotiations in the years preceding COP21 in 2015. However, discussions and decisions have largely focused on understanding and acknowledging the need for capacity building to promote climate action. COP21 in Paris saw the emergence of capacity building as a topic of debate for the first time, and a shift of attention towards delineating specific roles that the UN should play in capacity building. Delegates from the Umbrella Group, mostly consisting of non-EU industrial countries, argued that capacity building should continue to be led by

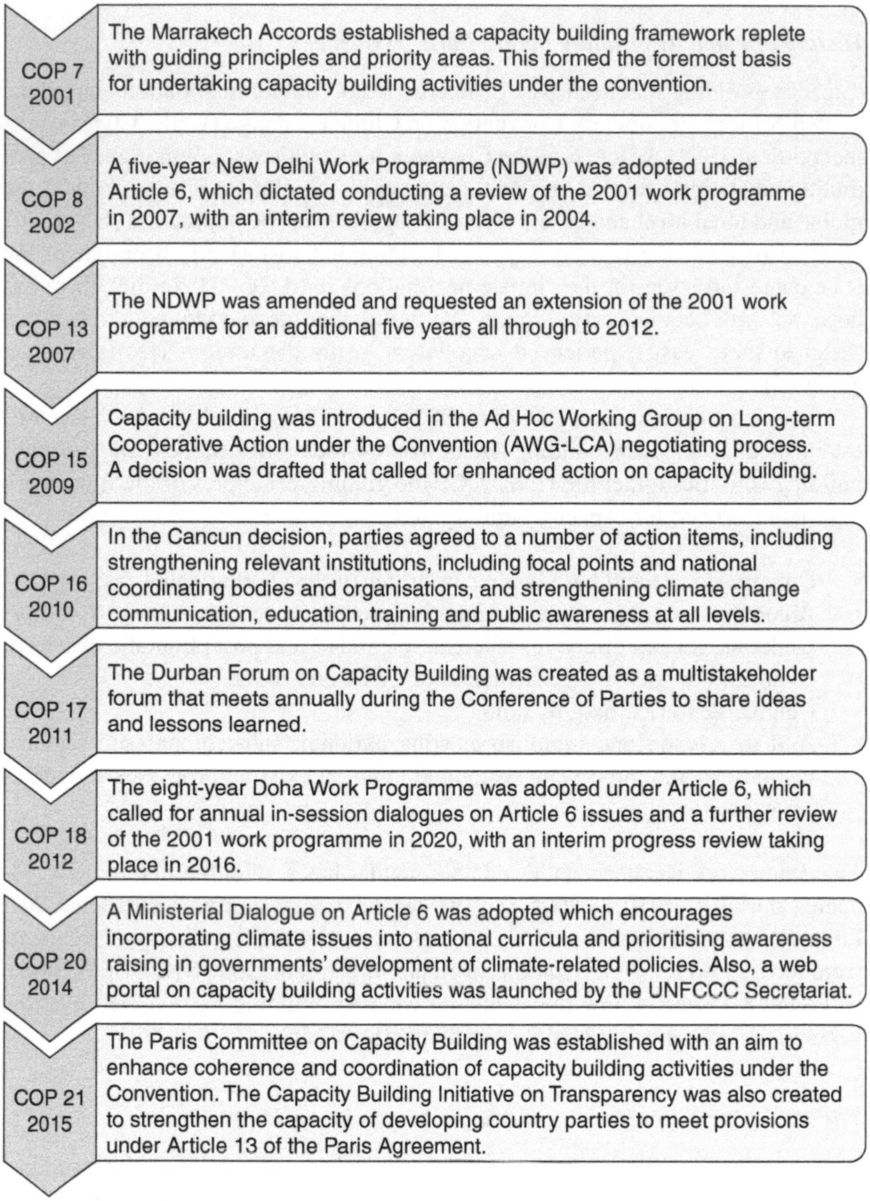

COP 7 **2001**	The Marrakech Accords established a capacity building framework replete with guiding principles and priority areas. This formed the foremost basis for undertaking capacity building activities under the convention.
COP 8 **2002**	A five-year New Delhi Work Programme (NDWP) was adopted under Article 6, which dictated conducting a review of the 2001 work programme in 2007, with an interim review taking place in 2004.
COP 13 **2007**	The NDWP was amended and requested an extension of the 2001 work programme for an additional five years all through to 2012.
COP 15 **2009**	Capacity building was introduced in the Ad Hoc Working Group on Long-term Cooperative Action under the Convention (AWG-LCA) negotiating process. A decision was drafted that called for enhanced action on capacity building.
COP 16 **2010**	In the Cancun decision, parties agreed to a number of action items, including strengthening relevant institutions, including focal points and national coordinating bodies and organisations, and strengthening climate change communication, education, training and public awareness at all levels.
COP 17 **2011**	The Durban Forum on Capacity Building was created as a multistakeholder forum that meets annually during the Conference of Parties to share ideas and lessons learned.
COP 18 **2012**	The eight-year Doha Work Programme was adopted under Article 6, which called for annual in-session dialogues on Article 6 issues and a further review of the 2001 work programme in 2020, with an interim progress review taking place in 2016.
COP 20 **2014**	A Ministerial Dialogue on Article 6 was adopted which encourages incorporating climate issues into national curricula and prioritising awareness raising in governments' development of climate-related policies. Also, a web portal on capacity building activities was launched by the UNFCCC Secretariat.
COP 21 **2015**	The Paris Committee on Capacity Building was established with an aim to enhance coherence and coordination of capacity building activities under the Convention. The Capacity Building Initiative on Transparency was also created to strengthen the capacity of developing country parties to meet provisions under Article 13 of the Paris Agreement.

Figure 8.1 Evolution of capacity building under UNFCCC.

their respective development agencies on an ad-hoc basis. Developing country parties, on the other hand, demanded that the UNFCCC play a more significant role to ensure coherence and coordination (Huq, 2016b).

This debate eventually led to recognition of capacity building in both the decision text from COP21 as well as the Paris Agreement. Article 11 of the Paris Agreement laid down the goals, guiding principles and procedural obligations of all the Parties to the Agreement with regard to capacity building. The Article stipulates that developed country Parties should support capacity building in developing countries whereas developing countries should regularly communicate progress made on implementing capacity-building plans, policies, actions and measures. In addition, Article 12 of the Agreement dictates that Parties cooperate in taking appropriate measures to enhance climate change education, training, public awareness, public participation and public access to information (Dagnet, Northrop, & Tirpak, 2015).

Paris Committee on capacity building

The decision text at COP21 established the Paris Committee on Capacity Building (PCCB), whose aim is to address both current and future capacity gaps and needs, as well as to further enhance existing capacity-building efforts by ensuring coordination and coherence among them.

COP21 also saw the launch of a workplan for the period 2016–2020, which the PCCB has been assigned to manage and oversee (UNFCCC, 2015).

Box 8.1 2016–2020 UNFCCC workplan on capacity building

The 2016–2020 workplan consists of nine components for furthering capacity-building activities under the UNFCCC.[1] These are:

a   Assessing how to increase synergies through cooperation and avoid duplication among existing bodies established under the Convention that implement capacity-building activities, including through collaborating with institutions under and outside the Convention;
b   Identifying capacity gaps and needs and recommending ways to address them;
c   Promoting the development and dissemination of tools and methodologies for the implementation of capacity building;
d   Fostering global, regional, national and subnational cooperation;
e   Identifying and collecting good practices, challenges, experiences and lessons learned from work on capacity building by bodies established under the Convention;
f   Exploring how developing country Parties can take ownership of building and maintaining capacity over time and space;
g   Identifying opportunities to strengthen capacity at the national, regional and subnational level;

---

h   Fostering dialogue, coordination, collaboration and coherence among relevant processes and initiatives under the Convention, including through exchanging information on capacity-building activities and strategies of bodies established under the Convention;

i   Providing guidance to the secretariat on the maintenance and further development of the web-based capacity-building portal.

j   https://unfccc.int/files/meetings/marrakech_nov_2016/application/pdf/auv_cop22_i13_3rd_review_of_cb_convention.pdf

Capacity-building initiative for transparency

The Paris Agreement under Article 13 also established the Capacity Building Initiative for Transparency (CBIT). The aim of the initiative is to strengthen institutional and technical capacity of developing country Parties in order to help them meet transparency requirements dictated under Article 13 (UNFCCC, 2015).

COP22 – from good intentions to sound action

With the Paris Agreement entering into force on 4 November 2016, COP22, held from 7–18 November 2016 in Marrakech, provided a timely opportunity to ensure that the agreement is put into practice. Negotiators focused on advancing work on capacity building, and technology development and transfer under the various mandates of the Paris Agreement. The 'Marrakech Action Proclamation for Our Climate and Sustainable Development' was issued, which calls for a greater volume, access to and flow of 'improved capacity and technology, including from developed to developing countries'.

The third comprehensive review of the capacity-building framework under the Convention was conducted at COP22, which invited the Paris Committee on Capacity Building (PCCB) to take the lead in managing and overseeing the 2016–2020 workplan on capacity building (UNFCCC, 2016c). The review also called for the following:

- Parties to foster networking and enhance their collaboration with academia and research centres, with a view to promoting individual, institutional and systemic capacity-building through education, training and public awareness.
- Subsidiary Body for Implementation (SBI) to facilitate complementarity between the Durban Forum and the Paris Committee on Capacity-building.
- Relevant intergovernmental and non-governmental organisations, as well as the private sector, academia and other stakeholders, to continue incorporating capacity building into their work programmes.
- United Nations agencies, multilateral organisations and relevant admitted observer organisations engaged in providing capacity-building support to developing countries to provide information to the secretariat to be uploaded on the capacity-building portal.

The conference also set aside a thematic day on Education to address the critical role of education in the global response to climate change, as acknowledged in Article 12 of the Paris Agreement. In addition, large volumes of financial support have also been pledged for promoting capacity building in developing countries. CBIT, which was established during COP21, was declared operational during climate talks at COP22 (GEF, 2016).

Tapping into the potential of universities in building climate capacity

Universities have historically been engaged in capacity building. Yet very few climate capacity-building initiatives to date have identified them as climate capacity-building hubs. This section will explore why and how universities could be engaged in climate capacity-building initiatives.

One of the rationales for considering universities is the fact that almost every country in the world has a university and these universities, whether located in the North or in the South, foster a culture of innovation and knowledge transfer. However, there are substantial capacity gaps between the universities of developed and developing countries. Nonetheless, university systems universally are built to assist local communities to climb knowledge ladders on issues such as finding smarter means to address climate change.

Another consideration is the fact that universities are among the oldest and most sustainable capacity-building institutions in the world. Universities can engage successive classes of students to study issues related to climate change. This is a continuous process which has the potential to transform expenditure into investments and could last 'for years after the funding is dispensed' (Hoff-meister et al., 2016). When these students eventually go on to become academicians, policy-makers and practitioners, they will carry forward the knowledge they gained and eventually contribute to mainstreaming climate change.

As mentioned earlier, Southern universities are yet to reach their full potential for building climate knowledge and capacity of students. Lack of basic resources including research funding, updated curricula, computing facilities, access to scientific publications and even internet are some of the contributing factors. Other barriers, such as language and economic conditions of students, also restrict Southern universities from reaching their full potentials. Furthermore, typically, very few developing country universities have research, training and teaching programmes targeting climate change. This is often due to the fact that very little climate capacity funding has been directed to universities.

To ensure effective participation of Northern and Southern universities alike in climate capacity building, it is necessary to enhance the capacity of Southern universities. Northern universities have a vital role to play in this and can do so without any substantial external funding. Potential ways to engage universities in climate capacity building more effectively are highlighted below.

Building stronger global alliances

Climate change, being a global problem, requires global partnerships for addressing the challenges. The Research and Independent Non-Governmental Organisations (RINGOs) are a constituency under the UNFCCC that is contributing to global climate capacity building. As the need for new information under the UNFCCC emerges, RINGO invites universities to conduct research on that particular issue. Apart from the RINGOs, as yet there has been little or no international initiative for enhancing collaboration amongst universities towards building climate capacity. While many universities strive to become engaged in global climate capacity building, they are often unaware of effective ways of doing so and efforts to build collaboration are often weakly formed or coordinated. At Marrakech, two specific initiatives for building alliances among universities were announced. The proposed networks have been named the Universities Network for Climate Capacity (UNCC) and the Least Developed Country University Consortium for Climate Change (LUCCC). While UNCC is envisioned to be a network of universities from all countries, LUCCC has been proposed initially for ten LDCs with a specific focus on climate change adaptation. How successful these two initiatives become in supporting global climate capacity building remains to be seen in the coming years. Lessons arising from the development and operation of the two networks will be immensely valuable for future initiatives that intend to engage universities in climate capacity building. Knowledge generated from such collaboration among universities should inform the activities of the PCCB, which in turn also needs to work as the key coordinator to match research needs with universities and other experts in order to build capacities in targeted areas.

Research without boundaries

Typically, research collaborations exist among universities. However, they are often limited within departments, universities and academia, leading mostly to joint publications or new projects. It is recognised that such scope and opportunities for conducting joint research is higher at universities in the developed world. On the other hand, researchers in developing countries have few resources and opportunities for collaboration among themselves, as well as with their counterparts from developed countries. Thus, new competencies and knowledge could be generated through robust capacity-building programmes that promote and support collaborations that go beyond borders, and also encourage university researchers (who are often criticised for working in silos) to work jointly with practitioners to solve complex real-life problems.

Box 8.2 LDC University Consortium on Climate Change (LUCCC)

Example of South–South collaboration on climate capacity building with a specific focus on adaptation, an area relatively less explored

During COP22 in Marrakech the International Centre for Climate Change and Development (ICCCAD) at Independent University, Bangladesh (IUB) held a side event at the University of Cadi Ayyad to discuss the role of universities in implementing Article 11 of the Paris Agreement. Another side event on the same topic was also held initiated by the Research and Independent NGOs (RINGOs) group, as well as ICCCAD at IUB and the UNFCCC Secretariat. Following overwhelming support from university students and faculty from least developed countries (LDCs), ICCCAD made a proposal to set up a consortium of universities to build climate capacity, with a specific focus on community-based adaptation. This proposed consortium, initially comprising 10 partner universities from LDCs, has been dubbed the LDC University Consortium on Climate Change (LUCCC). The partner countries are Bangladesh, Uganda, Tanzania, Bhutan, Ethiopia, Nepal, Mozambique, Gambia, Senegal and Sudan.

The key objective of the consortium would be to foster a South–South collaborative network among universities to help develop and implement teaching and training programmes on various climate change aspects, thereby enhancing research capacity and proficiency of LDCs on the issue. The focus of the consortium will be on adaptation; more specifically on community-based adaptation. The proposed activities are hosting faculty from partners and visiting researchers, sharing curricula and teaching tools, joint research and publications, scholarship for PhD and MSc students, exchange visits, and assisting the PCCB in taking informed decisions.

The consortium will be led and managed by two research organisations from LDCs: the International Centre for Climate Change and Development (ICCCAD) at the Independent University Bangladesh (IUB) and the Makerere University Center for Climate Change Research & Innovations (MUCCRI) in Uganda.

Enhanced access to information for enhancing mutual capacities

Limited access to information remains a major barrier for capacity building of Southern researchers and students. Universities in the developing countries lack access to the latest development information as they cannot support the costs associated with accessing databases, journals or professional development materials. Indeed, many developing country universities, especially in the least developed countries, have limited internet access, which further hinders capacity building. Northern universities can facilitate access to new knowledge and information to universities in the developing countries (e.g. access to thesis database, online libraries, etc.). While funding for climate capacity building is essential, universities could, through South–South, or South–North consortium, share resources to enhance mutual capacities.

Actions on climate change depend largely on the availability of sound and accurate data. For example, the scoping study on the National Mechanism on

Loss and Damage in Bangladesh identified the unavailability of data as one of the major challenges in introducing measures such as agricultural insurance. Regional cooperation among the universities can play a crucial role in accumulating and processing data and presenting them in a format useful to decision-makers and practitioners. Collaboration among researchers could also play a role in building capacity for data interpretation so as to assist in developing effective adaptation and mitigation measures.

Capacity development of the educators

Most universities in the Global North now offer stand-alone programmes or courses on climate change to their students. In contrast, many developing country universities are yet to update their curricula. Enhancing capacity of the educators at Southern institutions would enable them to design new climate change related courses and integrate materials into their classes in a useful way. Such knowledge sharing does not have to be unidirectional. Institutes in the North could also learn from their Southern counterparts (e.g. about various aspects of adapting to climate change) in order to enrich and diversify their content.

Distance learning and student exchange programmes can broaden the horizon

Online learning or distance-learning programmes can also be a means to enhance local capacity. Often the funds required to travel and participate in climate change related programmes and courses bar researchers and learners from doing so. However, universities could create opportunities for attending those courses online or via distance programmes, which could also serve the purpose to a great extent. Several Northern universities are already offering free courses that are available for online enrolment. Student exchange programmes could also enhance collaborations among universities to enhance climate capacity. Successful student exchange programmes often lead to valuable long-term relationships among universities. Universities can offer more climate capacity-building student exchange programmes without any additional financial support. For example, exchange students can work as interns on climate change research projects at a foreign school rather than taking classes for credit, and the student's home institutions could agree to award credit for the experience. The PCCB could collect information on such programmes and suggest ways to facilitate and improve student exchanges.

These are just some suggested ways for enhancing engagement of universities in climate capacity. Further thought should be given to designing more innovative programmes to enhance collaboration among universities. The PCCB could actually play a facilitating role in accomplishing the goal of engaging universities. During COP22 at a side event on 'The role of Universities in implementing Paris Agreement' at the University of Cadi Ayyad, participants suggested that a needs assessment be conducted for the capacity-building needs of the developing countries. Since under Article 11 of the Paris Agreement, the PCCB has the mandate to oversee all these capacity-building programmes and keep track of initiatives, they could potentially commission or convene the task. Also since Article 11 has guaranteed a place for sustainable capacity building under the UNFCCC, it can be expected that some level of investment will be made into long-term capacity building.

Understanding challenges within university systems

Despite the substantial promise, a number of challenges remain that may impede universities from effectively engaging in climate capacity building where most needed. Listed below are some of these potential structural challenges that often exist within university systems.[1]

- Disciplinary divisions within universities can hinder interdisciplinary research. Obstacles to effective climate action are as much scientific as they are sociological, political and economic. A truly interdisciplinary approach is, therefore, imperative in order to successfully identify and address all the challenges and solutions to the cross-cutting issue of climate change. However, universities are often characterised by disciplinary divisions that often have little coordination among them. This is a major impediment to furthering interdisciplinary research.
- Incentive structures at universities are not built to support applied, community-based research of the type that would be required for climate change adaptation and related capacity building. Incentive structures at universities usually place more weight on fundamental research rather than applied research such as on community-based climate change adaptation.
- University systems celebrate and cherish research publications. High emphasis is placed on publishing journal papers. The deliverable of capacity-building research would have to be more than just a research paper.
- Incentives for Northern universities to engage and invest resources in climate capacity building and also in assisting Southern universities should be identified. There is also lack of incentives for universities in developed countries to engage and invest resources in climate capacity building and in assisting Southern universities.
- Capacities of LDC universities need to be enhanced before they can engage in climate change capacity building. Universities in LDCs commonly suffer from a constraint of capacity, both technical and human. Before they can become effective catalysts in climate change capacity building, their own capacities need to be developed.
- Capacity-building partnerships need to be driven by strategies stemming from both Southern and Northern universities; Southern universities cannot just follow a package developed by others, which tends to be a common phenomenon. To facilitate climate action effectively, knowledge has to be drawn from their locale and be participatory.

• Universities are often criticised for high levels of bureaucracy, which can slow down progress. Also, as already mentioned, academics often work in silos and, unlike practitioners, often lack real world experience.

To become effective hubs for long-term climate capacity building, universities in both the South and the North would have to address these issues.

Thinking beyond the horizon

Capacity building is a 'fundamental precondition' for the post-2020 climate regime, which will require equal and active participation from all. While current practices of capacity building need to be redefined, and capacity gaps identified, politicians and development agencies will have to accept that a new paradigm of thinking is needed to bridge the capacity gap between the Global North and Global South, and long-term sustainable climate capacity-building initiatives would have to be undertaken. University systems define some of the most long-lasting capacity-building institutes globally and as such should be made an integral part of the long-term sustainable climate capacity-building paradigm. In the future, knowledge exchange will not be thought of as a Global North to Global South phenomenon: knowledge exchange and capacity building will be just as much a Global South to Global South experience, and even a Global South to Global North one. Greater collaboration among universities in the North and the South through sharing resources, and engaging in joint research programmes and student exchanges could assist universities from the South to build their own capacity and effectively contribute to capacity building. Universities have to be more innovative in collaborating with each other and at the same time they also have to think about addressing some of the structural challenges that currently exist within the university systems that may hinder their effective participation in sustainable climate capacity-building initiatives. The creation of the PCCB under the Paris Agreement is an important milestone towards revamping the climate capacity-building model. The PCCB could indeed play a greater role in ensuring that universities are more engaged in the process. The committee should also take initiative in restructuring the current capacity-building model and facilitate the development of scalable capacity-building programmes, instead of sporadic spending on presentations and seminars. 'Climate change is real, so should be capacity building' (Huq & Nasir, 2016).

Note

1 These challenges were highlighted by participants during a side event titled 'Role of Universities in Implementing the Paris Agreement', held during COP22 in Marrakech.

References

Chen, Z. & He, J. (2013). *Foreign aid for climate change related capacity building.* WIDER Working Paper 2013/46. Accessed on 28 January 2017 http://recom.wider. unu.edu/sites/default/files/Research%20brief%20-%20Foreign%20aid%2C%20 capacity%20building%20and%20climate%20change_1.pdf

Dagnet, Y., Northrop, E., & Tirpak, D. (2015). *How to strengthen institutional architecture for capacity building to support the post-2020 climate regime.* World Resources Institute.

GEF (2016). *New GEF fund gives boost to Paris Agreement implementation.* Global Environment Facility Press Release. Accessed on 26 January 2017 https://unfccc.int/ files/meetings/marrakech_nov_2016/application/pdf/gef_newfund_release_en.pdf

Hoffmeister, V., Averill, M., & Huq, S. (2016). *The role of universities in capacity building under the Paris agreement.* ICCCAD-CDL Policy Brief. Accessed on 28 January 2017. www.icccad.net/wp-content/uploads/2016/08/Capacity-Building-Policy-Brief-July-4.pdf

Huq, S. (2016a). Paradigm must shift for long term capacity building on climate adaptation. *Dhaka Tribune,* 19 November 2016. Accessed on 25 January 2017 www. dhakatribune.com/climate-change/2016/11/19/paradigm-shift-must-long-term-capacity-building/

Huq, S. (2016b). *Why universities, not consultants, should benefit from climate funds.* Climate Home, 17 May 2016. Accessed on 28 January 2017 www.climatechangenews. com/2016/05/17/why-universities-not-consultants-should-benefit-from-climate-funds/

Huq, S., & Nasir, N. (2016). *Stop sending climate consultants to poor countries – invest in universities instead.* The Conversation, 3 October 2016. Accessed on 24 January 2017 http://theconversation.com/stop-sending-climate-consultants-to-poor-countries-invest-in-universities-instcad-65135

Khan, M., Sagar, A., Huq, S., & Thiam, P. K. (2016). *Capacity building under the Paris Agreement. European Capacity Building Initiative.* European Capacity Building Initiative. Accessed on 28 January 2017. www.eurocapacity.org/downloads/Capacity_ Building_under_Paris_Agreement_2016.pdf

UNFCCC (1992). United Nations Framework Convention on Climate Change. Accessed on 28 January 2017. http://unfccc.int/files/essential_background/background_publications_ htmlpdf/application/pdf/conveng.pdf

UNFCCC (2015). Paris Agreement. Accessed on 28 January 2017. http://unfccc.int/files/ essential_background/convention/application/pdf/english_paris_agreement.pdf

UNFCCC (2016a). *Third comprehensive review of the implementation of the framework for capacity-building in developing countries.* Technical paper by UNFCCC secretariat. Accessed on 28 January 2017. http://unfccc.int/resource/docs/2016/tp/01.pdf

UNFCCC (2016b). *UNFCCC Capacity Building: Background.* Accessed on 26 January 2017. http://unfccc.int/cooperation_and_support/capacity_building/items/7061.php

UNFCCC (2016c), *Capacity-building: Paris Committee on Capacity-building.* Accessed on 26 January, 2017. http://unfccc.int/cooperation_and_support/capacity_building/ items/10053.php

Index

ABC model 64–6, 69–71; assumption of correspondence between attitudes and behaviours 68; economic-rationalistic 73

academia and private foundations 99–101, 103, 136, 138

Ad Hoc Working Group on Long-term Cooperative Action 134

Adams, John 17

adaptation 55, 78–9, 82, 84–6, 100, 111, 114–15, 132, 139; benefits of 85; community-based 139; developing effective 140; initiatives 92, 130; and mitigation 85; planning 83; strategies 91, 93, 95, 97, 99, 101, 103; and transformation 84

Adger, W. Neil 7–10, 76, 78, 80–2, 84–7, 90–1, 103

aerosols, sulphate 5, 113–17, 120–2

agency 2–3, 7–8, 31, 49, 63, 65, 82, 90, 98, 131; human 34; organized 55; political 7; public 81; and rationality 7; regulatory 109, 126; responses 81; social 5

agents 1–3, 39, 46–9, 51, 53, 55, 57, 65, 93, 115; collective 7, 47, 55; individual 7, 54, 65, 72–3; institutional 10, 55, 57; moral 48; rational 71; responsible 48; social 102

Agrawala, S. 9

air 110–11; ambient 111, 113; space 121

ambient air 111, 113

Antarctic ozone layer 4

anthropologists and anthropology 2, 8–9, 117

anti-Muslim sentiments 43

arguments 2, 16, 23–4, 34–6, 42–3, 52, 54, 63, 116, 131; anti-democratic 36; democratic 41; elitist 41; epistemic 39;

ethical 55; incompatibility of 116; moral 78; for proactionary policies 35–6; uncontested 39

Aristotle 47–9

Armeni, C. 121–2

attitudes 6, 65–6, 68, 70–1; and actions 69, 71; and actions of Danish people 68–9; environmental 66; individual 70; pro-environmental 63, 66; reactive 47

authoritarian 32, 36, 41; environmentalism 23, 27; forms of government 5, 23; proactionary policies 35; regimes 27, 36; states 23

BAE Systems (company) 120

Bangladesh 139–40

Beeson, M. 5, 23, 32

behaviour 5, 9, 49, 55, 57–8, 63–6, 68–73, 76, 78, 109; and choices 63; climate-smart 55; human 2–5, 38, 64–5; and incentive based approaches 64; of institutions 57–8; pro-environmental 65; responsible 58; restrictive 118; social 2

behavioural changes 9–10, 64, 85

behavioural patterns 64, 70

behavioural standards 93

behavioural studies 63, 65

Berlin, Isaiah 18

bibliometric studies 64

Bjurstro, M, A. 63–4

Blake, J. 69, 71

boroughs 91, 97, 101

boundaries 121, 124, 138

Brink, E. 93

building climate knowledge and capacity 17, 130, 137

Bulkeley, H. 8

Bush, George W. 118–19

Milton Keynes UK
Ingram Content Group UK Ltd.
UKHW040052071024
449327UK00019B/505